Crystallization of
Organic Compounds

Crystallization of Organic Compounds

An Industrial Perspective

Hsien-Hsin Tung
Edward L. Paul
Michael Midler
James A. McCauley

A JOHN WILEY & SONS, INC., PUBLICATION

Published by John Wiley & Sons, Inc., Hoboken, New Jersey
Published simultaneously in Canada

For general information on our other products and services or for technical support, please contact our Customer Care Department within the United States at (800) 762-2974, outside the United States at (317) 572-3993 or fax (317) 572-4002.

Wiley also publishes its books in a variety of electronic formats. Some content that appears in print may not be available in electronic formats. For more information about Wiley products, visit our web site at www.wiley.com.

Library of Congress Cataloging-in-Publication Data:

Crystallization of Organic Compounds: An Industrial Perspective/Hsien-Hsin Tung . . . [et al.].
    p. cm.
  Includes bibliographical references and index.
  ISBN 978-0-471-46780-9 (cloth)
  1. Crystallization--Industrial applications. 2. Pharmaceutical chemistry. 3. Pharmaceutical industry.
  I. Tung, Hsien-Hsin, 1955-TP156.C7I53 2009
  615'.19--dc22

2008042950

Printed in the United States of America

10 9 8 7 6 5 4 3 2 1

Contents

Preface

Crystallization is an essential operation in pharmaceutical manufacturing because the majority of active pharmaceutical ingredients (APIs) are produced in solid form. Yet, this subject is much less a part of the academic curriculum compared to other topics such as distillation, extraction, and reaction. Very often engineers will learn crystallization process development on the job through trial and error, and it is not surprising that wheels are reinvented from time to time, despite hard work and effort. In terms of resource utilization, this approach is certainly inefficient. Added to this deficiency is the lack of a mechanism to pass on the knowledge and expertise developed from previous efforts. Over the years, one way to accomplish this has been via memos and process reports. But memos are generally project specific. Therefore, it is not a trivial task to uncover the technical knowledge and know-how buried in various memos and reports. Combining a summary of relevant theory and illustrative examples in a book to fill this gap seems to be a good mechanism for the transfer of information on principles and suggested practice.

The idea of writing a book on crystallization to fulfill this need was first conceived in mid-1990. At that time, few books were available which dealt with crystallization development. These books appeared to overemphasize theory, and the majority of examples concerned crystallization of inorganic compounds. Over the past 10 years, several new crystallization books have been published which provide wider applications and richer information for development scientists and engineers. Unfortunately, the practical aspects of crystallization in our industries and actual industrial examples have not been adequately described.

This book has two goals. One is to facilitate the understanding of the fundamental properties of crystallization and the impact of these properties on crystallization process development. The second is to aid practitioners in problem-solving using actual industrial examples under real process constraints. This book begins with fundamental thermodynamic properties (Chapters 2 and 3), nucleation and crystal growth kinetics (Chapter 4), and process dynamics and scale-up considerations (Chapters 5 and 6). Subsequent chapters cover modes of crystallization operation: cooling (Chapter 7), evaporation (Chapter 8), antisolvent (Chapter 9), reaction (Chapter 10), and special cases of crystallization (Chapter 11). As mentioned, real industrial examples are provided in each chapter.

We would like to express our sincere thanks to the late Omar Davidson for his diligent support throughout the preparation of this book. We also want to thank our colleagues, Lou Crocker, Albert Epstein, Brian Johnson, Mamoud Kaba, Joe Kukura, Amar Mahajan, Jim Meyer, Russ Lander, Karen Larson, Chuck Orella, Cindy Starbuck, Jose Tabora, and Mike Thien, who have graciously spent their time in reviewing individual chapters of this book (and in several cases, more than that). Their recommendations have significantly enriched the content of this book. Needless to say, we are truly grateful to our spouses and family members for their understanding and support during the long period of preparation.

Our goal is to help reader develop the crystallization process. Matthew: **12**:33, "*Either declare the tree good and its fruit is good, or declare the tree rotten and its fruit is rotten, for a tree is known by its fruit.*" It is our hope that you, as readers, will find this book useful for your work. If so, this will be the nicest reward for us.

Chapter 1

Introduction to Crystallization Issues

Crystallization has been the most important separation and purification process in the pharmaceutical industry throughout its history. Many parallels exist in the fine chemicals industry as well. Over the past several decades the study of crystallization operations has taken on even higher levels of importance because of several critical factors that require increased control of the crystallization process. These levels of control require better understanding of the fundamentals as well as of the operating characteristics of crystallization equipment, including the critical issue of scale-up.

In the pharmaceutical industry, the issue of better control, desirable in itself, is reinforced by the need to assure the regulatory authorities that a continuing supply of active pharmaceutical ingredients (APIs) of high and reproducible quality and bioavailability can be delivered for formulation and finally to the patient. The "product image" (properties, purity, etc.) of this medicine must be the same as that used in the clinical testing carried out to prove the product's place in the therapeutic marketplace. Some additional comments on regulatory issues are included later in this chapter (Section 1.7).

The issues noted above that require increased control, relative to previous practice, include the following:

- Final bulk drug substances must be purified to high levels that are increasingly quantifiable by new and/or improved analytical methods.
- Physical attributes of the bulk drug substance must be better controlled to meet formulation needs for reproducibility and bioavailability.
- Many APIs now require high levels of chirality.
- Increased demands are being made for achievement and maintenance of morphology.
- Increasingly complex molecular structures with higher molecular weights are being processed.
- Bulk drug solid stability is increasingly being achieved by improved control of crystal growth.
- The biotechnology sector has increased the use of precipitation of macromolecules for purification and isolation of noncrystalline materials.

Added to this list is the assertion, based on operating experience, that crystallization is difficult to scale up without experiencing changes in physical attributes and impurity

Crystallization of Organic Compounds: An Industrial Perspective. By H.-H. Tung, E. L. Paul, M. Midler, and J. A. McCauley

rejection. Regulatory requirements for final bulk drug substances, as noted above, now include the necessity for duplication of physical attributes including particle size distribution, bulk density, and surface area within narrow ranges when scaling from pilot plant to manufacturing scale.

When compared to the development of models and methods for other unit operations, it is obvious that crystallization has not been generalized to the degree that has been accomplished for distillation, extraction, adsorption, etc. This situation is changing rapidly, however, with increasing research now being carried out at academic and industrial centers on crystallization fundamentals to model and predict nucleation and/or growth rates as well as other key properties, including polymorph formation.

Control of crystallization processes requires modulation of either nucleation or growth or, as is most often the case, both modes of crystal development simultaneously. Each operation must be evaluated to determine which of these process objectives is most critical, from the point of view of overall outcome, to determine whether nucleation or growth should be the dominant phase. Much of the literature is focused on nucleation for the obvious reason that the number and size of nuclei initially formed can dominate the remainder of the operation. However, it is generally agreed that nucleation can be difficult to control, since there are several factors that can play a role in the conditions for nucleation onset, nucleation rate, and number of crystals generated before growth predominates.

The demand for increasing control of physical attributes for final bulk pharmaceuticals has necessitated a shift in emphasis from control of nucleation to control of growth. This trend is also finding application for control of purity and improved downstream handling for both intermediates and final bulk products. The obvious critical factors then become *seeding and control of supersaturation.* Quantification of these factors for each growth process is essential for development of a scalable process. Much of the discussion to follow focuses on the growth process and methods to minimize nucleation.

The purpose of this book is to outline the challenges that must be met and the methods that have been and continue to be developed to meet these requirements to develop reproducible crystallization operations and to design equipment with which these goals can be achieved.

The four conventional crystallization operations (see Chapters 7, 8, 9, 10) will be discussed in terms of their strengths and weaknesses in achieving specific process objectives. In addition, methods of augmenting the conventional processing methods will be considered, with emphasis on the enhanced control that is often necessary to achieve the specific objectives.

This book also includes chapters on the properties of organic compounds (Chapter 2), polymorphism (Chapter 3) and the kinetics of crystallization (Chapter 4), critical issues (Chapter 5), and mixing effects in crystallization (Chapter 6). Chapter 11 includes areas of current crystallization research and development we thought worth mentioning and also some unique crystallization processes that have special features to be considered in process development. To assist in the thought process for organization of a new crystallization process, Chapter 11 also contains a suggested protocol for development and scale-up of a crystallization operation.

1.1 CRYSTAL PROPERTIES AND POLYMORPHISM (CHAPTERS 2 AND 3)

Basic crystal properties include solubility, supersaturation, metastable zone width, oil, amorphous solid, polymorphism, occlusion, morphology, and particle size distribution. Clearly.

in order to properly design and optimize crystallization processes, it is essential to have a sound understanding of these properties.

For pharmaceuticals and special organic chemicals, solution crystallization, in which solvents are used, is the primary method of crystallization compared to other crystallization techniques such as melt or supercritical crystallization. Therefore, the goal of these chapters is to introduce basic properties of solution and crystals related to solution crystallization. The relevance of these basic properties to crystal qualities and crystallization operations will be highlighted with specific examples.

Some properties are more clearly defined than others. For example, solubility is defined as the amount of solid in equilibrium with the solvent. Solubility can affect the capacity of the crystallization process, as well as its ability to reject undesired compounds and minimize loss in the mother liquor. In addition, solubility varies widely from compound to compound or solvent to solvent. On the other hand, there are properties that are much less well characterized or understood. For example, the mechanism and condition for the formation of oil or amorphous solid remain unclear. The composition of oil and amorphous solid can be variable, and certainly can contain a much higher level of impurities than that in the crystalline solid, which leads to a real purification challenge. In addition, oil or amorphous solid generally is less stable and can create critical issues in drug formulation and storage stability.

One property of a crystalline compound is its ability to form polymorphs, that is, more than one crystal form for the same molecular entity. The phenomenon of polymorphism plays a critical role in the pharmaceutical industry because it affects every phase of drug development, from initial drug discovery to final clinical evaluation, including patent protection and competition in the market. A critical challenge is the early identification of possible polymorphs. Chapters 2 and 3 will address this key issue.

1.2 NUCLEATION AND GROWTH KINETICS (CHAPTER 4)

Meeting crystal product specifications with a robust, repeatable process requires careful control and balancing of nucleation and growth kinetics. Careful structuring of the environment can dictate the fundamental mechanisms of nucleation and crystal growth and their resultant kinetics. Undesired polymorphs can be often minimized or eliminated by suitable control of rate processes.

Understanding of the possible nucleation and crystal growth kinetics for desired (and undesired) compounds can place the process development effort on a considerably shorter path to success. Reference will be made to examples in the other chapters in this book.

1.3 CRITICAL ISSUES (CHAPTER 5)

Difficulty in controlling crystallization processes in general can be exacerbated when working with complex organic compounds. This problem can be even worse when attempting to develop a nucleation-dominated process, which, even in the best circumstances, can potentially operate over a very wide range of supersaturation, depending on small changes such as varying amounts of very-low-level impurities.

Organic compounds are subject to agglomeration/aggregation effects and, even worse, to "oiling out." All of these problems can potentially result in undesired trapping of solvent and/or impurities in the final crystal. Oiling out, of course, can completely inhibit the formation of a crystalline phase, resulting in a gum or an amorphous solid. These phenomena are discussed qualitatively in Chapter 5.

Crystalline processes often provide a seed bed for crystal growth with an initial nucleation step. When attempting to control particle size and shape, an excessive number of nuclei can effectively make it impossible to achieve the desired size or morphology. Optimal processes with externally or internally (heel) added seed often require some level of seed conditioning. Principles for such conditioning are discussed in Chapter 5 and in some of the examples.

Instrumentation for control of seed point and growth/nucleation processing is discussed.

1.4 MIXING AND CRYSTALLIZATION (CHAPTER 6)

While many crystallization processes can tolerate a wide range of mixing quality and intensity, many engaged in development do not examine the effect of mixing on their process until forced to do so by problems in scale-up or even possibly at laboratory scale. The result is, at best, loss of time and effort.

Transport of momentum, mass, and energy, all affected by mixing, can be critical for success in many crystallization processes, especially with complex organic compounds. Momentum transport can influence slurry homogeneity, impact nucleation, shear damage, agglomerate formation, and discharge of slurry. Mass transport can affect the uniformity of supersaturation (micro-, meso-, and macromixing), and in reactive crystallization can affect, even at the molecular level, the resultant reaction and subsequent supersaturation pattern. Energy transport has a direct effect on heat transfer, and proper mixing can minimize or avoid encrustation on the heat transfer surfaces.

An adaptation of the Damkoehler number (Da) is a useful concept for evaluation of mixing effects in crystallization. It is the ratio of the characteristic mixing time to its corresponding process time (nucleation induction time, crystal growth/supersaturation release time, or reaction time). Studies of these times and the resulting predicted Damkoehler number in a laboratory setting can provide evidence of possible scale-up problems.

The effects of mixing on surface films in crystal growth, and on mixing/local homogeneity when adding antisolvent or reagent, are examined in Chapter 6. Low-shear options (impeller design, vessel geometry—e.g. fluidized bed, contoured bottom) are also discussed.

1.5 CRYSTALLIZATION PROCESS OPTIONS (CHAPTERS 7–10)

The following is a qualitative discussion of several of the procedures that are used to create and maintain conditions under which crystallization can be carried out. These procedures create supersaturation by different methods and utilize seeding to varying degrees. The procedures are classified by the manner in which supersaturation is generated.

The equally critical issues of when to seed and how much seed to use are introduced in each classification. The amount of seed can vary from none to massive and include the familiar classifications of "pinch" to hopefully avoid complete nucleation, "small" (<1%) to hopefully achieve some growth, "large" (5–10%) to improve the probability of growth, and "massive" (the seed is the product in a continuous or semicontinuous operation) to provide maximum opportunity for all growth. The amount of seed can also be critical in the control of polymorphs and hydration/solvation.

The important and developing methods of online measurement of solution concentration and particle size and count are adding powerful tools to aid in the control of crystallization operations both in experimentation and manufacturing operations (Nagy et al. 2007). These methods will also be discussed in the context of their utilization.

1.5.1 Cooling (Chapter 7)

1.5.1.1 Batch Operation

Cooling a solution from above its solubility temperature can be performed in a variety of ways, depending on the system and the criticality of the desired result. Natural cooling, as determined by the heat transfer capability of the crystallizer, is the simplest method but results in varying supersaturation as the cooling proceeds. This may or may not be detrimental to the process, depending on the nucleation and growth rate characteristics of the particular system. Natural cooling has the potential to decrease the temperature rapidly enough to pass through the metastable region and reach the uncontrolled nucleation region before seeding can be effective. Uncontrolled nucleation can be a major problem with the potential to cause oiling out, agglomeration and/or fine particles, a larger particle size distribution (PSD), and occlusion of solvent and impurities. A secondary disadvantage of uncontrolled cooling can be accumulation of crystal scale on the cooling surface caused by low temperatures at the wall. Accumulation of a scale layer can be triggered by nucleation on the cold surface followed by growth on the thickening scale. This encrustation can severely limit the cooling rate, as well as cause major issues of nonuniformity in the product.

When high supersaturation is not acceptable, cooling strategies can be utilized to match the cooling rate with the increasing surface area. These rates were derived by Mullin and Nyvlt (1971) and further derived by Mullin (1993) and are very useful in control of supersaturation. They prescribe cooling rates that are much slower at the outset than natural cooling in order to maintain supersaturation in or close to the growth region when the crystal surface area for growth is low. The cooling rate can be increased as the surface area increases. An added benefit of this method is the potential to reduce encrustation by limiting temperature differences across the jacket. In theory, encrustation can be eliminated if the temperature difference between the cooling fluid and the crystallizing mixture is less than the width of the metastable zone (Mersmann 2001, pp. 437 ff.).

A further refinement of this strategy is described by Jones and Mullin (1974), in which a seed age is added as a further aid in limiting the development of supersaturation, thereby reducing nucleation and promoting growth.

Another key variable in batch cooling is seeding. The difficulty is in determining the seed point, which is ideally when the batch temperature first crosses the saturation curve. However, this temperature can be affected by batch-to-batch variations in several factors, including the actual concentration of the material to be crystallized, as well as by impurities that can affect the solubility. If the seed is added at a temperature above the solubility temperature, some or all of it can dissolve, resulting in uncontrolled nucleation. If the seed is added at a temperature too far below saturation, the product may have already nucleated. In either case, the increase in nucleation could result in a decrease in impurity rejection and/or a change in particle size distribution and other physical attributes.

This issue, determining the point of seeding, is common to crystallization by cooling, as well as solvent removal by concentration, and by antisolvent addition. As such, seed point determination merits discussion of various methods.

Online, in-situ instrumentation to measure product composition has been developed to successfully determine the seed point, and is being utilized in an increasing number of crystallization operations. Image analysis or photographic methods may be useful in determining the presence of nuclei >5 microns but would be too late to determine the point of seeding. These methods can be used, however, to determine if seeding was successful and to observe whether or not excessive nucleation has occurred. Incorporation of an age period at constant temperature after seeding can also help normalize the nucleation/growth ratio.

Adding the seed as slurry in the proper solvent composition is one of the best methods to control a batch cooled crystallizer. The slurry addition is started before reaching saturation and is continued until it can be determined that the seed is no longer dissolving. Although this method can increase the probability that seed will be present at the start of crystallization, the amount of seed actually remaining may be subject to excessive variation.

Crystallization by cooling may not be feasible when polymorphs are stable at different temperatures within the cooling range (Saranteas et al. 2005). Cooling through these regions of stability can result in mixed morphologies or a change from one polymorph to another. Uncontrolled nucleation can also be a major issue in achieving a uniform product when polymorphs are possible. A constant-temperature process with either a high level of seed or massive seed may be required to select the desired polymorph. Hydrates and solvates may also be subject to these factors in crystallization processes. Polymorphism is the subject of Chapter 3.

1.5.1.2 *Continuous Operation*

The batch-to-batch variation discussed above for batch cooling methods can be largely overcome by utilizing continuous operation to achieve both control of low levels of super-sturation and operation with massive amounts of seed. This technology is widely used for high-volume products but finds less application in the pharmaceutical industry because of lower volumes and campaigned operations in which continuous operations are more difficult to justify. However, in some examples discussed below, there is no alternative to continuous operation to achieve the separation and purification required.

A primary example is the resolution of optical isomers by continuous crystallization in fluid beds. Control of low supersaturation by control of the temperature difference between the continuous feed and the seed bed is critical to maintaining an essentially all-growth regime in which the individual isomers grow on their respective seeds in separate crystallizers. The seed beds in both crystallizers are massive in relation to the amount of racemic solution passing through in order to present sufficient seed area to maintain low supersaturation. Uncrystallized isomers in the overhead streams are recycled to dissolve additional racemic feed. Crystal size is maintained by sonication. See Examples 7-6 and 11-6 for a discussion of resolution of optical isomers by continuous crystallization.

This special case illustrates the power of continuous cooling processes with massive amounts of seed to reject impurities that have the potential to crystallize at equilibrium. Batch cooling to achieve this separation of optical isomers is not a practical alternative because the resolution is not based on equilibrium solubility. The time required for batch cooling would result in the nucleation of the undesired isomer when any practical amount of product is to be harvested in each cycle.

A high degree of control can also be achieved in continuously stirred tank crystallizers. Temperature differences between feed and crystallizer can be regulated as necessary. The seed is the product and will normally be present at the slurry concentration as determined by the feed rate, concentration, and solubility differences achieved. However, in cases in which this amount of seed is not sufficient, cross-flow filtration on the discharge of the crystallizer(s) can be used to increase the slurry density. See Example 7-4 for a discussion of the resolution of ibuprofen lysinate.

1.5.2 Concentration of Solvent (Chapter 8)

1.5.2.1 Semibatch Operation

Increasing the concentration by removing solvent by evaporation (semibatch operation) is widely practiced but has several nucleation and growth control problems. These problems can be sufficiently severe to make this method unsuitable in some cases, such as for final bulk drug substances (API) that may require tighter control of mean particle size and PSD than can be achieved on scale-up.

Evaporation rate is analogous to cooling rate in creating supersaturation and may be controlled by similar methods of control to match-evaporation rate with the surface area available for growth. The point of seeding is also an issue since it is difficult to determine when the saturation line is being crossed as concentration increases. Adding the seed as slurry in the evaporation solvent as the concentration passes through saturation can be useful in this regard.

Local variation in supersaturation is the most significant control issue that can cause non-reproducibility in PSD and other physical attributes, as well as solvent and impurity occlusion. These local variations occur both at the heating surface and at the boiling liquid/vapor interface.

At the heating surface, local high temperatures and a high vaporization rate result in uncontrollable local supersaturation environments in which uncontrolled nucleation can be excessive, particularly in those regions of poor bulk mixing. Wall scale above the heated surface can also lead to significant product quality issues. Decomposition on the surface above the liquid–vapor interface can be excessive because of direct exposure to the higher temperature of the heating fluid. Product scale from this area could also drop into the product slurry and result in unacceptable physical properties for a final bulk drug substance as well as handling difficulties in any system. Finally, overconcentration can lead to safety issues if the concentrated mass is thermally unstable. Although this is not a crystallization issue, it is mentioned as a possible serious consequence of an evaporative crystallization operation.

At the boiling surface, vapor disengagement can lead to very high local supersaturation as well as nucleation induced by vapor–liquid interfaces. Foaming can also be a significant issue. In addition, throughout the bulk, vapor bubbles can cause local nucleation.

These sources of variability all contribute to potentially severe scale-up problems with evaporative crystallization. Control of the distillation rate by control of the jacket temperature may require higher wall temperatures, thereby making supersaturation variation more severe. The decrease in bulk circulation and the increase in mixing time will further exacerbate this problem. In some cases, these problems can produce unacceptable results, requiring development of an alternative crystallization method. See Example 8-2 for a discussion of an application in which adequate PSD control could not be achieved.

1.5.2.2 Continuous Evaporation

Although widely practiced for production of industrial chemicals, continuous evaporation for crystallization is rarely if ever used in pharmaceutical operations. Although continuous operation has the advantages of using massive seeding and increased control of supersaturation and the crystal surface area, the throughput necessary for its application is rarely, if ever, achieved for final bulk drug substances. In addition, continuous operation to achieve the conditions for crystallization (as discussed above for resolution of optical isomers) is often not applicable or achievable. Local supersaturation at the liquid–vapor–solid interfaces is the primary cause of uncontrolled nucleation.

1.5.3 Addition of Antisolvent (Chapter 9)

1.5.3.1 Semibatch Operation

This widely used procedure has many inherent potential advantages over both batch cooling and concentration in terms of crystallization control. It does, however, have the obvious disadvantage of creating solvent mixtures requiring separation for recovery.

Control of both supersaturation and crystal growth area is readily achievable by control of the antisolvent addition rate. This control requires consideration of both the change in solubility as addition proceeds and the crystal growth area and is, therefore, potentially more complex than for the single-solvent processes of cooling and concentration. Rates of anti-solvent addition can vary from constant in noncritical cases to "cubic" (as in cooling operations), depending on the slope of the saturation curve with concentration. Solubility curves of unusual shape, possibly including a maximum over the range of addition, may require a more complex addition scheme if maintenance of essentially constant supersaturation in the metastable region is necessary.

Determination of the seed point is again the key to consistent operation. Addition of the anti-solvent containing seed during the segment in which the saturation line is crossed is a good method of seed control. Massive seeding is also possible by utilizing a significant portion of the previous batch as the seed.

Scale-up of these processes requires careful consideration of the mixing of the antisolvent, both at the point of addition and in circulation of the bulk. Insufficient control of local mixing at the point of addition can result in local supersaturation and excessive nucleation. Subsurface addition of the antisolvent is a good precaution to minimize this risk and is, in some cases, essential for successful scale-up. Micro- and macromixing issues in crystallization have been analyzed by Mersmann and Kind (1988) and Mersmann (2001, p. 418). Overmixing is also an issue since shear can break crystals and create nuclei by secondary nucleation. Rasmussen (2001) has devised a loop reactor/crystallizer for separately evaluating the effects of macro-, micro-, and mesomixing. Designed for reactive crystallization, this loop design can also be used to assist in scale-up of antisolvent crystallization processes. These issues are further discussed in Chapter 6 on mixing effects.

1.5.3.2 Semicontinuous Antisolvent Addition

Excellent control of crystallization conditions can be achieved by semicontinuous methods in which the supersaturation is controlled locally at the point of mixing in an in-line device. Both once-through and recycle operations can be carried out with and without seeding. In the case of unseeded operation, an in-line device can create a high supersaturation ratio in a very short time and provide a method of control of nucleation that is difficult or impossible to achieve in conventional crystallization vessels.

1.5.3.3 Impinging Jet Crystallization

The rapid blending of two steams that is achieved by impinging jet technology, as developed for reaction injection molding by Edwards (1984) was adapted for crystallization by Midler et al. (1994) and further developed by others [examples: Mahajan and Kirwan (1996),

Lindrud et al. (2001), Johnson and Prud'homme (2003)]. With proper design, mixing to the molecular level can be accomplished in less time than the nucleation time, thereby achieving a primarily nucleation-based process for the production of uniform, fine particles. After the nuclei leave the mixing zone, additional crystallization continues in a standard agitated vessel on a well-defined initial number of nuclei with a well-defined size and shape.

This technology can produce narrow particle size distributions with a controlled surface area and is finding utilization for final bulk drug substances. Control of particle size has the added benefit of eliminating the need for milling for particle size reduction and control. In addition, scale-up can be achieved to production scale by operation at the same local conditions in the same (or only approximately two times larger) size jets that are run for longer times. See Examples 9-5 and 9-6.

1.5.4 Reactive Crystallization (Chapter 10)

When supersaturation of a crystallizing compound is created by its formation by chemical reaction, the operation is characterized as reactive crystallization. The reaction may be between two complex organic compounds or can be neutralization by an acid or base to form a salt of a complex compound. These reactions can be very fast compared to both the mass transfer rates to the crystals, and the growth rate of the crystals, thereby leading to high local supersaturation and nucleation. These operations are also known as precipitations because of the rapid inherent kinetics.

Control of particle size in reactive crystallization is difficult because there is usually no practical method to slow down the reaction that generates the supersaturation. The rate of addition of the reagents, however, does provide a means to control this critical parameter globally in the reactor but not locally since the reaction may be complete near the point of addition. Successful operation depends, therefore, on a careful balance between addition rate of the reagent(s), local supersaturation, global supersaturation, mass transfer, and crystal growth surface area. Controlled supersaturation at the initiation of addition of the reagent(s) requires an initial charge of seed to prevent uncontrolled nucleation and the resulting creation of an excess number of particles. The seed must be developed in a separate operation because the intrinsic reaction may only generate crystals that are too small to be used as seed if a basic growth process is required. See Examples 10-1, 10-2, and 10-3.

1.6 SPECIAL APPLICATIONS (CHAPTER 11)

This chapter includes a discussion of several special topics on crystallization, including ultrasound for crystallization, crystallization using supercritical fluids, and experimental design and process control. It also contains examples of crystallization operations that were developed to meet special requirements.

The use of ultrasound in crystallization can be unique and very helpful in certain applications. In Examples 7-3 (heel/sonication), 7-6, and 11-6 (stereoisomer resolution), ultrasound was used to break up crystals and generate a fresh surface area for subsequent crystal growth. In these cases, the crystals were snapped into shorter crystals along the long axis. Therefore, the aspect ratio was effectively reduced. Improving the crystal aspect ratio by breaking up crystals along the long axis and/or by facilitating growth on the slowest-growing surface can be very useful in applications involving needles, as discussed in Examples 7-6 and 10-1.

One key driver for supercritical crystallization is generation of nanoparticles for improvement of the drug dissolution rate (Gupta 2006; York 2004). Fundamentally, this approach is very similar to impinging jet crystallization, as shown in Example 9-5, where high supersaturation is generated by mixing two streams rapidly. Several types of supercritical crystallization operation have been developed successfully to multikilogram scale.

With the advancement of online measurement techniques such as focused beam reflectance measurement (FBRM) and Fourier transform infrared (FTIR), it is now possible to obtain particle size distribution and solution concentration information rapidly through these in-situ probes. In one experiment, hundreds of data points can be generated. With proper experiment design, the model-based experimental design for crystallization is capable of obtaining high-quality crystallization kinetic data with a small number of experiments. This approach can thus save significant experimental effort and time in the development of crystallization processes.

Computational fluid dynamics (CFD) is increasingly being utilized to analyze mixing systems, particularly the stirred vessels commonly used for crystallizer operation (Woo et al. 2006). The problem of modeling fluid dynamics in the presence of a solid phase is not trivial, but some workers are starting to make headway in this field. These efforts are referred in Chapter 11.

Examples in this chapter include sterile crystallization of a labile compound, yield enhancement by crystallization, yield and selectivity enhancement, removal of low-level impurities via crystallization from the melt, crystal formation in vials in a freeze drier, and non-equilibrium resolution of stereoisomers by crystallization. These examples represent unique crystallization processes designed for specific purposes. One lesson to be learned from examination of these nonmainstream applications is that understanding of principles can lead to inventive solutions to problems. For instance, in Examples 11-2 and 11-3, the solubility difference between starting material and desired product is used to optimize the reaction yield/selectivity by crystallizing the product and protecting it from overreaction.

It should be noted that development of the crystallization processes in most of the examples presented in later chapters occurred before the availability of many of the online measurement and control methods that are now available. Utilization of these methods would have aided both the process development and the manufacturing operations. The literature that describes these methods—for example, feedback control of supersaturation for crystallization (Nonoyama et al. 2006; Zhou et al. 2006)—is now extensive, and the instrumentation to carry out the measurements and control continues to be improved.

1.7 REGULATORY ISSUES

Controlled crystallization methods and equipment are required not only to meet internal standards, such as consistency for intermediates and particularly for active APIs, but also to meet regulatory requirements. These requirements include controls on both chemical purity and physical attributes.

For APIs, limits are set on chemical purity, mean particle size, PSD, and other appropriate physical attributes by the biobatch model for clinical evaluation. The term "biobatch" refers to the regulatory requirement of identifying a particular batch, normally a pilot scale batch used in clinical trials, as the defining standard for physical and chemical attributes that must be reproduced at the manufacturing scale to be acceptable for sale. The critical process attributes (CPAs), once established, must be met on scale-up to the manufacturing facility. In addition, the process must be operated within the ranges established as critical process parameters (CPPs). Development of a crystallization process must include determination of realistic and reproducible ranges for both the CPPs and the CPAs.

One of the most difficult processes to scale up successfully is crystallization. Methods to achieve control of nucleation and growth are keys to development, and the degree to which they are successfully applied can be the difference between success and failure on scale-up. It is to this fundamental problem that this book is addressed, combining critically important teachings from the literature with personal experience of the authors and their colleagues in a variety of crystallization operations.

Chapter 2

Properties

For pharmaceuticals and special fine organic chemicals, solution crystallization, in which solvents are used, is the primary method of crystallization, in comparison to other crystallization techniques such as melt or supercritical crystallization. The goal of this chapter is to introduce basic properties of solution and crystals in order to better understand, design, and optimize the crystallization processes. The relevance of these basic properties to crystal qualities and crystallization operations will be highlighted, accompanied with specific examples.

With regard to specific properties, we will focus on those that are more relevant to solution crystallization and those that have a direct impact on the quality of the final bulk pharmaceuticals such as purity, form, habit, and size, based upon our own experience. We will leave readers to find other properties, such as miller index for crystal morphology, hardness of crystals, interfacial tension, etc., in other books on crystallization (Mersmann 2001; Mullin 2001), which provide in-depth theoretical discussion on these properties.

2.1 SOLUBILITY

Understanding the solubility behavior is an indispensable requirement for the successful development of crystallization processes. It is necessary to know how much solute can dissolve in the solvent system initially and how much solute will remains in the solvent at the end in order to conduct the crystallization operation. For solution crystallization, solubility of a chemical compound is simply the equilibrium (maximum) amount of this compound that can dissolve in a specific solvent system. A solution is saturated if the solute concentration is at its solubility limit. At saturation, no more dissolution will occur and the concentration of dissolved solute in the solution remains unchanged. Hence, the solid solute and dissolved solute are at equilibrium under this condition.

2.1.1 Free Energy–Composition Phase Diagram

In order to deepen our understanding of the thermodynamic feature of solubility, we present Fig. 2-1, a free energy–composition phase diagram (Balzhiser et al. 1972, pp. 437–443).

Figure 2-1 shows the Gibbs free energy profile of a binary system, where G on the y-axis represents the Gibbs free energy of the system and X on the x-axis represents the mole fraction of the desired compound. Specifically, the first component in this system can be the desired compound, and the second component can simply be the solvent. Both temperature and pressure are maintained constant in this case.

Crystallization of Organic Compounds: An Industrial Perspective. By H.-H. Tung, E. L. Paul, M. Midler, and J. A. McCauley
Copyright © 2009 John Wiley & Sons, Inc.

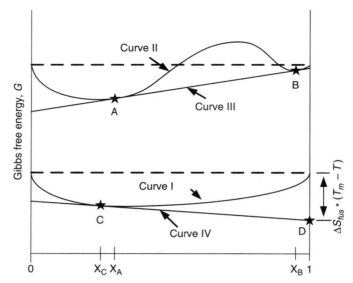

Figure 2-1 Free energy profile as a function of system composition and the conditions for the formation of two equilibrium phases.

In Fig. 2-1, curve I behaves as an upward concave curve over the entire range of composition. In this scenario, the binary mixture at any composition has a lower Gibbs free energy than two individual components. Therefore, the binary mixture at any composition is more stable than two individual components and will become a single phase. Mathematically, the shape of the concavity of this curve can be described as

$$G < 0 \quad \text{and} \quad \frac{\partial^2 G}{(\partial x)^2} > 0.$$

Another scenario involving curve II contains two upward concave curves and one downward concave curve. In addition, the third curve III, is tangent to curve II at points A and B of the lower part of two upward concave sections. The Gibbs free energy at points A and B can thus be expressed as a linear combination of the two intercepts of the tangent line on the y-axis at the composition of 0 and 1. Therefore, the intercept of this tangent line on the y-axis is equivalent to the partial molar Gibbs free energy of component one (the desired compound) and component two (the solvent) at a composition of X_A or X_B. Furthermore, for component one, the partial Gibbs free energy at composition X_A is identical to its partial Gibbs free energy at composition X_B. This situation also applies to component two. In other words, points A and B are at equilibrium. Point A represents the first equilibrium (solvent with dissolved drug) phase, and point B represents the second equilibrium oil or amorphous drug containing solvent phase. Points A and B are called the binodal points in the free energy composition diagram. The binodal points A and B can be functions of temperature and pressure. If the system's composition lies below point A, it is undersaturated. If the system's composition is between points A and B, it is supersaturated. We will discuss the supersaturated condition later in Section 2.2.1.

Analogously, for solid–liquid equilibrium, it can be expressed as curve IV tangent to curve I at point C (equilibrium solubility) and intercepting the y-axis at point D (equilibrium crystalline solid) for the compound of interest. Mathematically, it can be written as follows (Reid et al. 1977, pp. 380–384):

$$\mu \, (\text{solid phase}) = \mu \, (\text{liquid phase})$$
$$= \mu \, (\text{pure compound as liquid at temperature } T) + RT \cdot \ln a \quad (2\text{-}1)$$

or equivalently

$$\ln a = \ln x_i^{SAT} \cdot \gamma_i^{SAT} = \frac{\Delta_{fus}S}{R} \left(1 - \frac{T_m}{T} \right) \quad (2\text{-}2)$$

where μ is the partial molar Gibbs free energy, T is temperature, T_m is the melting point of the compound, R is the Boltzman constant, ΔS is entropy of fusion, a is activity of the compound of interest in solution, which is directly related to the amount of compound dissolved, i.e., solubility x_i^{SAT}, and activity coefficient γ_i^{SAT}. In Equation 2-2, it is assumed that the difference in heat capacity of the compound as a liquid and as a solid is negligible. The readers can find a more detailed discussion on phase equilibrium of multicomponent systems in the above references.

The purpose of these two equations is to show that temperature, difference in the chemical potential of a compound as a solid and a liquid, or entropy of melting can directly affect solubility. In addition, solvent, as well as impurities in the solution, can affect the activity coefficient. The chemical structure and salt forms of the compound can affect the entropy of melting and of the activity coefficient, and hence solubility. In the following sections, we will elucidate the impact of these variables on solubility from a practical point of view.

2.1.2 Temperature

The impact of temperature on solubility is demonstrated in Fig. 2-2. This figure shows the solubility profile of lovastatin, a cholesterol-lowering drug, in a methanol/water mixture as a function of temperature. As shown in Fig. 2-2, solubility increases as temperature increases. This solubility behavior is commonly observed in organic compounds and agrees quite well with Equation 2-2.

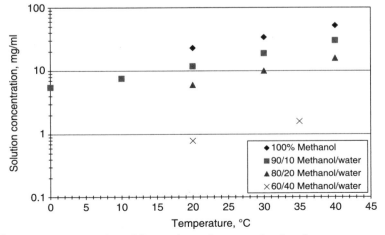

Figure 2-2   Solubility of lovastatin in different solvent mixtures as a function of temperature.

In certain compounds, especially inorganic salts, solubility may not be affected or may actually decrease at higher temperatures. A good example is calcium carbonate water (hardwater), which has lower solubility at higher temperatures. This reverse solubility behavior causes troublesome scale deposits issue in hot water boilers because water is maintained at a higher temperature in the water boiler. Therefore, calcium carbonate becomes supersaturated and precipitates in the boiler. In the authors' experience, however, reverse solubility is very rare in fine organics or pharmaceuticals.

Because temperature can have a strong influence on solubility, it is a commonly used to control the crystallization operation. The impure material can simply dissolve in a particular solvent system at an elevated temperature, and the pure material can crystallize from the solution by lowering the temperature.

2.1.3 Solvent

The solvent plays a critical role in altering solubility as well. Figure 2-3 shows the solubility profile of lovastatin in a methanol/water solvent system at room temperature. Water is used as an antisolvent in this particular example. As shown in Fig. 2-3, solubility decreases sharply as water percentage increases. This is a typical behavior in which solubility is reduced monotonically as the antisolvent percentage increases.

The behavior of solubility may be highly nonlinear in certain solvent systems. Figure 2-4 shows the solubility profile of a mesylate intermediate of a drug candidate in a solvent mixture of toluene/acetonitrile. Acetonitrile is used as an antisolvent in this particular example. As shown in Fig. 2-4, the solubility curve reaches a maximum at the midpoint of the solve mixture and decreases at higher or lower percentages of acetonitrile. This behavior is less frequent than the monotonic behavior shown in Fig. 2-3, but it still occurs. A simple analogy of this nonmonotonic behavior to vapor–liquid equilibrium would be the existence of (low-boiling point) azeotrope, which has a lower boiling point than either pure component. This reflects the impact of the solvent on the activity coefficient of the solute and changing its solubility, as shown in Equation 2-1.

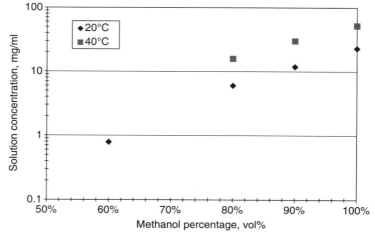

Figure 2-3   Solubility of lovastatin as a function of solvent composition.

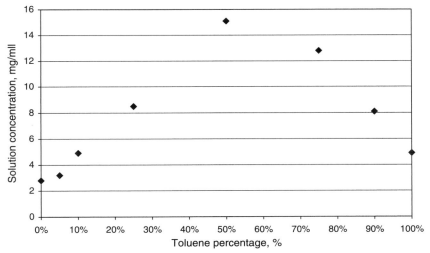

Figure 2-4   Atypical solubility behavior: a drug intermediate as a function of solvent composition. Solubility reaches a maximum at a certain solvent composition.

Like temperature, solvent composition is a common variable in controlling the crystallization process. The impure material can be dissolved in a particular solvent system, and the pure material can be crystallized from the solution by adding the antisolvent.

Solvent can strongly affect other crystallization variables, including polymorphism, solvate, crystal morphology, crystallization kinetics, etc. The impact of solvent on these variables will be addressed throughout the book as appropriate.

2.1.4 Impurities

Impurities can also influence solubility to a significant extent. Figure 2-5 shows the solubility of lovastatin in pure solvent and in mother liquors. Mother liquor refers to the supernatant

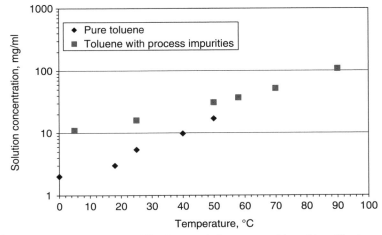

Figure 2-5   Impact of impurities on solubility. In general, the presence of impurities will enhance solubility.

of the slurry after crystallization. Mother liquor generally contains impurities rejected during crystallization. As shown in Fig. 2-5, lovastatin's solubility is significantly greater in mother liquors than in pure solvent.

Solubility enhancement in the presence of impurities, especially in the mother liquor, is a familiar phenomenon. Impurities in the mother liquor may not be well characterized, although their chemical structures have some similarity to the desired compound. In practical applications, the impact of impurities on solubility generally is unknown a priori and must be determined experimentally. Due to the potential impact of impurities on solubility, care should be taken in conducting crystallization experiments if the starting materials have varying levels of impurities from batch to batch. The presence of impurities can further affect crystallization kinetics, which will be addressed in the next chapter.

2.1.5   Chemical Structure and Salt Form

If two compounds have similar chemical structures, we tend to assume that they have similar solubilities. However, we should be cautious in making this assumption. Figure 2-6 shows the solubilities of lovastatin and simvastatin in a methanol/water solvent system. Despite the fact that simvastatin has only one extra methyl group, the solubilities of the two compound are significantly different.

If a compound contains an (or multiple) acidic or basic functional group, forming a salt can significantly alter its solubility. Needless to say, different types of salts can have quite different solubilities.

Since varying the salt form can affect the solubility significantly, this is another useful technique in conducting crystallization. For example, the desired compound may have low solubility in the selected solvent. However, after forming the salt, its solubility can increase significantly and become completely solvable. To crystallize the desired pure salt, either cooling or adding antisolvent can be done. Alternatively, the desired compound can dissolve in the selected solvent and be converted to salt that has low solubility in this particular

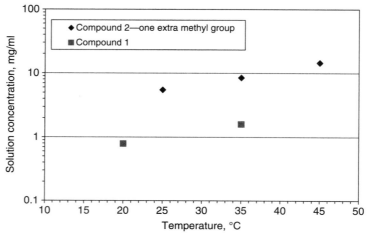

Figure 2-6   Impact of chemical structure on the solubility of lovastatin (compound 1) and simvastatin (compound 2) with one extra methyl group.

solvent. Therefore, the salt will crystallize from the solution—an approach typically called reactive crystallization. In practice, both methods have been used.

Like varying the solvent, varying the salt form also affects other crystal properties including morphology, polymorphs, crystallization kinetics, formulation, drug stability, etc. Therefore, selection of the salt form always involves other considerations in addition to solubility.

2.1.6 Solubility Measurement and Prediction

In the laboratory, individual solubility measurement can be performed quite easily with a simple setup. A solvent or solvent mixture and an excessive amount of solid are charged into a jacketed glass flask. The slurry is agitated and maintained at the desired temperature for a certain period, preferably overnight. Then a sample of slurry is taken and filtered through a filter. The clear filtrate is diluted for a high-performance liquid chromatography (HPLC) assay. Generally, sampling and filtering of the slurry are conducted at room temperature for ease of handling. This can introduce errors of measurement if the temperature at which solubility is measured is not the ambient temperature. Many modifications of this protocol are possible. For example, a filter can be incorporated into the solubility measurement glass flask so that in-situ filtration can be performed.

Given the limited material supply in the early phase of drug development and the need to identify the proper solvent and temperature range for crystallization as quickly as possible, high-throughput screening devices, which originated in the combinatorial approach to drug discovery (Lipinski et al. 2001), are getting much wider usage. Many commercial units are currently available on the market. In general, the high-throughput screening device contains multiple cells or vials from microliter to milliliter range, which enable users to conduct multiple experiments with a very limited supply of materials. The key advance of this technology lies in automation. All operations are automated so that multiple samples can be handled repetitively (Symyx and Crystal) (Alsenz and Kansy 2007).

Another trend in industry is the use of in-situ analytical instruments, such as FTIR, Raman, ultraviolet/visible (UV/Vis) or near infrared (NIR) spectrophotometers, which are part of the process analytical technologies (PATs) (Dunuwila and Berglund 1997; Podkulski 1997; Togkalidou et al. 2002). These types of instruments typically contain a sensor probe which can be immersed directly in the process stream or batch. Therefore, they avoid (offline) sampling completely and all the complications associated with sampling. In comparison to the high-throughput screening approach, PAT is more suitable for process monitoring rather than analytical analysis.

Spectrophotometers can be used to measure the solubility or solution concentration. In order to establish the correlation between solubility/solution concentration and spectra peak areas, calibration with known concentrations is required. However, as with any technology, there are limitations and issues associated with in-situ analytical instrumentation, such as minimum detection level, peak shift, peak overlapping, etc. Development of an accurate correlation model for specific compounds over a wide range of temperature and solvent composition may still require a fair amount of effort.

Prediction of solubility is also receiving more attention, mostly limited to academic research. In comparison to the vapor-liquid equilibrium situation, which has built an extensive database for reliable prediction (Reid et al. 1977), prediction of solid-liquid equilibrium remains in its early stage (Kolar et al. 2002). However, this field is developing rapidly, and its future potential cannot be overlooked (Tung et al. 2007).

2.1.7 Significance for Crystallization

Solubility is a key element in the development of a crystallization process. As mentioned earlier, solubility allows us to determine how much solute can dissolve in the solvent system initially, how much solute will remain in the solvent at the end of crystallization, and how much of each of the impurities, if their solubilities are available, can be rejected.

For example, if the solubility of a desired compound is 100 mg/ml at 50°C and 10 mg/ml at 0°C, it is feasible to design a crystallization process which dissolves 100 gm of desired compound at 50°C and crystallizes 90 gm of compound at 0°C per liter of solvent. The yield of this crystallization process is

$$90 \text{ gm}/100 \text{ gm} = 90\%.$$

Solubility ratios of the desirable compound over the undesirable impurities and the purity of starting material limit the maximum theoretical yield achievable with acceptable purity for the desirable compound. Continuing the previous example, let us assume that the crude starting material contains 15 wt% of one key impurity and the solubility of this impurity is identical to the solubility of the desirable compound. If we charge 100 gm of crude solid, which has 85 gm of desired compound and 15 gm of impurity, into 1 liter of solvent, the final crystallized product will contain

$$85 - 10 = 75 \text{ gm of desired compound}$$

and

$$15 - 10 = 5 \text{ gm of impurity.}$$

So, the final product has a purity of

$$75/(75 + 5) = 93.75 \text{ wt\%.}$$

For this case, the yield will be

$$75/85 = 88\%.$$

To improve the yield, we can increase the initial crude charge to 117 gm of crude per liter of solvent so that it has approximate 100 gm of desired compound. The yield of this approach will increase to its maximum yield as

$$(100 - 10)/100 = 90\%.$$

But the purity of the final product will drop to

$$90/[(17 - 10) + 90] = 92.78\%.$$

The purpose of the above calculation is to illustrate the impact of solubility on the design of the crystallization process. Since it is not always possible to measure the solubility of impurities, the relationship between yield and product purity may not be determined a priori. But the general relationship between yield and product purity will still hold: the higher the yield, the lower the purity, and vice versa. Solubility information is clearly useful to determine the limitation of crystallization performance.

Solubility information is also required in determining the supersaturation and crystallization kinetics. These subjects will be discussed in more detail in the following sections and the next chapter. In addition, solubility information may be used to optimize the overall synthesis of pharmaceuticals. In Chapter 11 readers will find novel applications, Examples 11-2 and 11-3, which take advantage of the solubility difference between starting materials and final product to improve the reaction selectivity, overall yield, and product purity and reduce the raw material cost.

2.2 SUPERSATURATION, METASTABLE ZONE, AND INDUCTION TIME

The solution is supersaturated when the solute concentration exceeds its solubility limit. A solution may maintain its supersaturation over a concentration range for a certain period without the formation of a secondary phase. This region is called the metastable zone. From the creation of supersaturation to the first appearance of the secondary (solid) phase, the time elapsed is called induction time. As supersaturation increases, the induction time is reduced. When the supersaturation reaches a certain level, the formation of the secondary phase becomes spontaneous as soon as supersaturation is generated. This point is defined as the metastable zone width. Figure 2-7 is a typical diagram of the equilibrium solubility curve and the metastable zone curve (Mullin 2001).

2.2.1   Free Energy–Composition Phase Diagram

A supersaturated solution is always located beyond its solubility line, just as a saturated solution is always located at the equilibrium line. In order to provide more fundamental insight into the metastable phenomenon, a free energy–composition diagram, which provides further detail than is shown in Fig. 2-1, is presented in Fig. 2-8. The reader can find more detailed discussion in the following references (Balzhiser et al. 1972; Debenedetti 1995).

As shown in Figs. 2-1 and 2-8, points A and B are equilibrium points at the lower part of two upward concave sections on curve II. In Fig. 2-8, two additional points C and D along curve II are identified. These are inflection points where the upward concave curve turns into

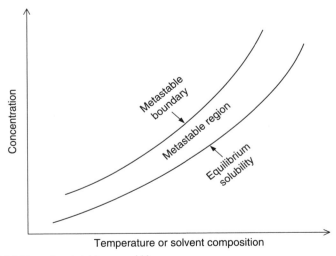

Figure 2-7   Solubility and metastable zone width.

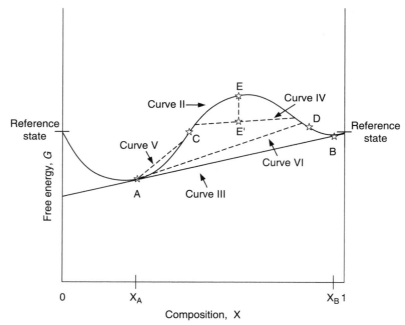

Figure 2-8   Qualitative illustration of the relationship of the free energy profile and the metastable zone width. Beyond the metastable zone width, any disturbance to the system will result in a mixture which has a lower free energy than in the initial condition. Within the metastable zone width, the system could be metastable and remain supersaturated, or it can form a second phase with certain disturbances.

the downward concave curve. Points C and D are called the spinodal points and can be expressed mathematically as

$$\frac{\partial^2 G}{(\partial x)^2} = 0.$$

Spinodal points are unstable points on the free energy–composition phase curve. A system with a composition within the region of C and D will split spontaneously into two phases and eventually reach points A and B. For example, for point E within this region, it can split into two phases connected through curve IV, which has the same material composition as point E. The summation of the free energy of these two phases is equal to point E′, which is always below the original free energy of point E. Hence the region between points C and D is unconditionally unstable. We may view spinodal point C as the metastable zone width of crystallization.

For any point within the metastable region, i.e., between points A and C, a small perturbation to the original state may result in a split of the two phases shown as curve V. As shown in Fig. 2-8, curve V has a higher Gibbs free energy than the original state. Therefore, the phase split is unfavorable. The system will return to its original state and remain supersaturated. However, if there is sufficient perturbation to the system so that the phase split can cross over to the other side of free energy–composition phase curve such as curve VI, the decomposition results in a lower overall Gibbs free energy than the original state. For this scenario, the split is favorable. The system will release its supersaturation and the secondary phase will appear.

Clearly, depending upon the nature of the system, a supersaturated solution could have a wide range of metastable zone width. Also, the supersaturated solution may remain metastable for a long time, i.e., a long induction time, before it forms the secondary solid phase.

2.2.2 Factors Affecting Metastable Zone Width and Induction Time

Similar to solubility, the metastable zone width and induction time of a supersaturated solution are affected by various factors, including temperature, solvent composition, chemical structure, salt form, impurities in the solution, etc. Therefore, although the spinodal point is a thermodynamic property, it is very difficult to measure the absolute value of the metastable zone width experimentally. Regardless, understanding the qualitative behavior of the metastable zone width and the induction time can be helpful for the design of crystallization processes.

From the point of view of free energy–composition phase diagram, it is reasonable to expect that the distance between the spinodal and binodal points should become smaller as the temperature of the system increases if there are no additional complications such as thermal or chemical decomposition. This expectation is based upon the assumption that at sufficiently high temperatures, two phases will mix freely with each other (assuming totally miscibility). Therefore, the system becomes a single phase, i.e., curve II converges to curve I or points A and B move closer and merge with each other, as shown in Fig. 2-1. If this happens, it means equivalently that the metastable zone also becomes *narrower* and the induction time is *shorter* at higher temperatures. Broadly speaking, this assumption is consistent with the authors' experience with cooling crystallization. For antisolvent or reactive crystallization, it seems reasonable to expect that the metastable zone width will be narrower and the induction time will be shorter at a lower antisolvent level. However, more data are needed to substantiate this hypothesis. As mentioned earlier in Section 2.1.3, there could be many exceptions to this hypothesis. As for the impact of impurities, as the level of impurities in the solution increases, the metastable zone and induction time generally become broader and longer.

Other factors, such as the presence of seed of the desired compound, undissolved extraneous solid particles, and even agitation intensity can affect the metastable zone width and induction time. Clearly, these factors can perturb and alter the free energy–composition phase curve. In general, these factors will lower both the metastable zone width and the induction time.

2.2.3 Measurement and Prediction

Determination of the solution concentration, either at supersaturated or saturated states, can be done offline by taking slurry samples from the process, similar to the measurement of solubility in Section 2.1.6. This method involves complications due to temperature variation and solvent evaporation. In addition, sampling is generally time and labor intensive. As mentioned earlier, in-situ analytical instrument such as FTIR or UV-visible spectrophotometry can measure the solution concentration and calculate the supersaturation given the solubility information. Accurate determination of supersaturation can greatly increase the understanding of crystallization kinetics and the development of the crystallization process.

Reliable determination of metastable zone width and induction time-generally is more time-consuming and difficult than the determination of supersaturation. This is because metastable zone width and induction time are affected by various factors. Therefore, the data may not be reproducible when the operating environment is changed. On the other hand, accurate measurement of metastable zone width or induction time is not critical for many industrial applications. A few data points for rough estimation are generally sufficient for the design of crystallization processes.

To measure the metastable zone width, the first step is to prepare a slightly undersaturated solution based upon solubility data. For cooling crystallization, the clear undersaturated solution is cooled at a finite rate to generate the supersaturation. During cooling, the solution becomes cloudy and marks the boundary of the metastable stable zone at the particular cooling rate. Since the cooling rate can affect the measured metastable zone width, this procedure may need to be repeated several times using different cooling rates. The asymptotic point with a zero cooling rate defines the ultimate true metastable point. To construct the entire metastable zone width curve, this approach will be repeated at different initial solution concentrations. For antisolvent crystallization, similar protocols can be developed to measure the metastable zone width. The procedure for measuring the metastable zone width can be facilitated by using an in-situ particle size analyzer, such as the focussed beam reflectance measurement (FBRM) device that can detect and measure the particle size without sampling for external analysis (Birch et al. 2005).

The discussion of solubility above shows that it is theoretically possible to deduce the metastable zone width through the free energy–composition curve if such a curve is available (Kim and Mersmann 2001). Due to the complexity of the problem, experimental verification is preferred for practical applications.

2.2.4 Significance for Crystallization

The ability to generate supersaturation is a requirement for crystallization. As inferred from Section 2.1, supersaturation can be created by varying the temperature, the solvent composition, and its salt form. Practical cases based upon the method of generation of supersaturation will be presented in detail in Chapters 7 to 10.

In general, for batch crystallization with a narrower metastable zone width, the operating window for generation of supersaturation is smaller. It is more likely to create nucleation with fine crystals, and vice versa.

It is not always desirable to have a wide metastable zone width, though. Figure 2-9 shows the concentration profile of a tartrate salt during the course of crystallization. The tartrate salt was formed by the addition of L-tartaric acid to a solution of free base in a solvent mixture of acetonitrile/ethyl acetate/water. As shown in Fig. 2-9, the solution remains highly supersaturated (metastable) over an extended period after the addition of tartaric acid to the free base if the seed surface area is not sufficiently high. The presence of a highly supersaturated metastable solution, coupled with a slow release of supersaturation, can be problematic. First, slow release of supersaturation will significantly lengthen the batch time cycle. Second, the highly supersaturated solution may lead to unpredictable nucleation, which would be difficult to control and which would create in inconsistent performance from laboratory to factory and from batch to batch.

There are also cases that tentatively exceed the metastable zone width during crystallization—for instance, impinging jet crystallization, shown in Examples 9-5 and 9-6. Since the induction time becomes extremely short, the final product size depends strongly on the distribution of local supersaturation. Successful process design to obtain a consistent particle size distribution will require tight control of mixing and generation of supersaturation. As mentioned, a detailed description of impinging jet crystallization is given in Examples 9-5 and 9-6.

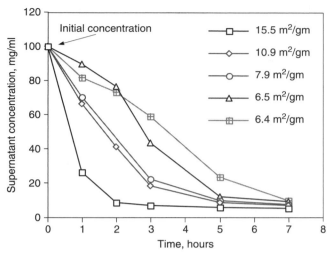

Figure 2-9   Solution concentration profile during crystallization. This figure shows the impact of seed surface area on the release of supersaturation. If the seed surface area is insufficient, the system will become highly supersaturated during crystallization.

Metastability and induction time can play an interesting role in the kinetic resolution of optical isomers, such as of resolution of ibuprofen lysinate (see Example 7.4). On the one hand in order to maintain the optical purity of the desired isomer, it is necessary to keep the (undesired) isomer in its supersaturated state for the entire crystallization period. If the undesired isomer crystallizes out from the solution, the optical purity of the desired isomer will decrease. On the other hand, it is important to release the supersaturation of the desired isomer, which starts at the same initial level of supersaturation as the undesirable isomer. These are two conflicting requirements. To overcome this dilemma, it is critical to maintain a large amount of seed bed of the desired isomer to accelerate the release of super-saturation of the desired isomer, whereas the undesired isomer remains supersaturated. As mentioned, a detailed description of the resolution process of optical isomers is given in Example 7-4.

2.3   OIL, AMORPHOUS, AND CRYSTALLINE STATES

Upon release of supersaturation, the initially dissolved compound will be separated from the solution and form a secondary phase, which could be either oil, amorphous solid, or crystalline solid. Crystalline materials are solids in which molecules are arranged in a periodical three-dimensional pattern. Amorphous materials are solids in which molecules do not have a periodical three-dimensional pattern. Under some circumstances with very high supersaturation, the initial secondary phase could be a liquid phase, i.e., oil, in which molecules could be randomly arranged in three-dimensional patterns and have much higher mobility than solids. Generally, the oil phase is unstable and will convert to amorphous material and/or a crystalline solid over time. At a lower degree of supersaturation, an amorphous solid can be generated. Like the oil, the amorphous solid is unstable and can transform into a crystalline solid over time. Even as a crystalline solid, there could be different solid states with different crystal structures and stability. The formation of different crystalline solid states is the key subject of polymorphism, which will be mentioned below and discussed in more detail in Chapter 3. The transformation from unstable states to stable states is generally called Ostwald Rule of the Stages effect (Mullin 2001).

The oil, amorphous, and crystalline states of materials are phenomena frequently encountered during crystallization. In the following discussion, we will give examples of these phenomena based upon their dependence on supersaturation.

2.3.1 Phase Diagram

Figure 2-10 outlines the relationship of oil, amorphous material, and crystalline material with respect to supersaturation. We should emphasize that this diagram is primarily based upon empirical observation over years of development of crystallization with various compounds. The authors do not intend to use it to build a theoretical framework. In comparison to meta-stable zone width, there is relatively little discussion on the oiling phenomenon in the literature (Bonnet et al. 2002; Lafferrere et al. 2002). Therefore, it is beneficial to present such a diagram even without much theoretical derivation.

As shown in Fig. 2-10, when an extremely highly supersaturated solution is created, it can form a secondary oil (and/or gel) phase initially. As an example, Fig. 2-11 shows the appearance of oil droplets in a highly supersaturated simvastatin solution. In this example, the secondary phase of oil droplets was observed initially after mixing a clear, concentrated batch solution with enough antisolvent to form instantaneously a highly supersaturated solution. The resulting oil droplets were unstable and quickly transformed into small crystalline solids upon aging. It has also been observed that in other compounds oil droplets convert to amorphous solids upon aging.

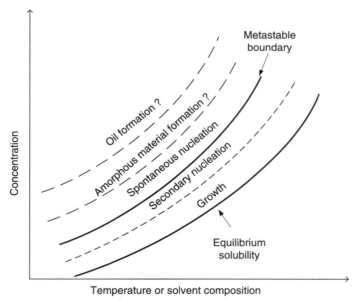

Figure 2-10   A hypothetical expanded map between supersaturation and various crystallization phenomena. When the system is highly supersaturated, it may form oil droplets, an amorphous solid, or nuclei from spontaneous nucleation. When the system is less supersaturated, it may form nuclei from secondary nucleation or just growth on the existing crystal surface.

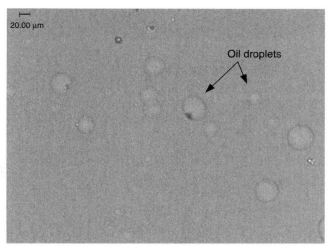

Figure 2-11   Formation of oil droplets of a highly supersaturated API solution.

For a less supersaturated solution, Fig. 2-12 indicates that the amorphous solid can be generated after aging of the oil droplets. An amorphous solid can also be formed by rapid cooling of a saturated solution, as in lyophilization or freeze-drying operations. Like oil droplets, amorphous solids are unstable and will transform into crystalline material upon aging under proper conditions. As shown in Example 12-5, the amorphous imipenem can be transformed into crystalline material even under freezing conditions.

The most stable and desirable solid phase upon release of supersaturation is clearly a crystalline solid, as shown in Fig. 2-13. A crystalline solid is generally purer and more stable than an amorphous solid and oil droplets. The majority of organic chemicals, especially pharmaceuticals, are produced as crystalline solids.

We should point out that in practice we may observe all three states. In simvastatin as shown in Fig. 2-11, under the microscope, the oil droplets seem to convert directly to a

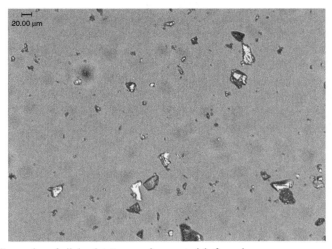

Figure 2-12   Conversion of oil droplets to amorphous material after aging.

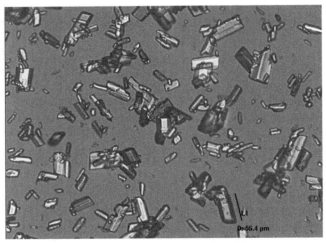

Figure 2-13   A crystalline solid formed under less supersaturated conditions.

crystalline solid without the intermediate amorphous solid. For some materials, we have never observed the formation of oil droplets under the conditions of crystallization. However, the generality of Fig. 2-11 appears to be valid for a wide range of compounds and has been proven useful in guiding us through the development of crystallization.

2.3.2 Measurement

Oil and amorphous or crystalline materials can be easily distinguished under the microscope. Oil is simply a secondary liquid phase. Amorphous material generally has no definitive shape with no birefringence under polarized light. Crystalline materials typically have a defined shape, showing birefringence under polarized light (Stoiber and Morse 1994).

For more quantitative measurement, X-ray powder diffraction analysis is commonly used in industry (Hancock 1997; Jenkins 2000). A crystalline solid will typically show several distinct peaks which reflect the unique structure of crystals, whereas an amorphous solid will have no distinct peaks. This technique could be further used to assess quantitatively the degree of amorphism within a solid, as mentioned in Example 12-5 (Connolly et al. 1996). In this case, it has been demonstrated that the shift of baseline of X-ray diffraction spectra is proportional to the degree of amorphism in the solid phase (Crocker and McCauley 1995).

Another frequently used technique is differential scanning calorimetry. When a solid sample is heated at a preselected heating rate, the resulting heat flow profile over temperature reveals the degree of crystallinity of the solid sample. A 100% crystalline solid should melt only at or near its melting point. Therefore, the heat flow profile should be smooth over the entire heating period and will only detect an endothermic heat flow corresponding to enthalpy of melting at its melting point. If some amorphous solid material is present in the solid, it is likely to have some degree of solid-state transition and corresponding heat flow prior to melting. In the study of simvastatin, it is shown that differential scanning calorimetry can be applied to differentiate the defects of crystals generated by several different crystallization methods (Elder 1990).

2.3.3 Significance for Crystallization

Oil droplets generally contain very high levels of impurities, including solvents, and create very difficult purification problems. In addition, the oil phase is essentially a liquid phase and cannot be isolated as a solid for further drug formulation work. The oil phase is unstable and will likely convert into an amorphous or crystalline solid upon aging, but without much control of the final particle size. Therefore, the formation of the oil phase becomes a diaster for the process. Seeding and control of supersaturation are common techniques to overcome this difficulty (Deneau 2005). Amorphous material has complications of impurity rejection and solid state instability similar to those of the oil material. Therefore, the formation of an amorphous solid is generally undesirable, although sometimes the compounds cannot generate a crystal form. Certain drugs, in particular those that are produced via freeze drying, are inherently amorphous solids. These amorphous drugs are made primarily to improve their dissolution rate in the human body, but at the expense of drug stability and subambient storage temperature (Leuner and Dressman 2000). The crystalline solid remains by far the most desirable form for drug formulation, although more attention has recently been given to amorphous solid dispersion (Lowinger et al. 2007).

A more detailed discussion on the impact of the phenomena described in Fig. 2-11 on the process development of crystallization is given in Chapter 5.

Maintaining crystallinity is required not just during crystallization, but also in subsequent steps such as the filtration, drying, and milling. A hygroscopic solid will absorb water and may lose its crystallinity (see Section 2.9). Therefore, controlling drying and storage humidity is critical for this type of material. Some crystalline materials may have tendency to lose their crystallinity with drying (heat) or milling (pressure). Special attention is required to maintain crystallinity.

2.4 POLYMORPHISM

A compound may form crystalline solids which have different three-dimensional structures. This results in polymorphs (Brittain 1999). With this definition, all polymorphic crystals should be the same molecular species, but with different crystal structures. Note that optical isomers, which have the same chemical structure, are considered different molecular species and should not be confused with polymorphism. Only a brief description of this subject is given below. A more detailed discussion is provided in Chapter 3.

2.4.1 Phase Diagram

Polymorphism crystals typically exhibit different physical properties and behaviors, such as different solubilities and melting points. Figure 2-14 shows solubility curves of monotropic and enantiotropic polymorphs. For the monotropic polymorphs, polymorph I has lower solubility than polymorph II throughout the entire solubility range. For the enantiotropic polymorphs, polymorph I has higher solubility than polymorph III at the lower part of solubility range but lower solubility that polymorph III at the higher part of solubility range, or vice versa.

Polymorphism is a common phenomenon in the chemical and pharmaceutical industries. Figure 2-15 illustrates the solubility of Forms I and III polymorphs of a reverse transcriptase inhibitor drug candidate which exhibit enantiotropic behavior. Figure 2-16

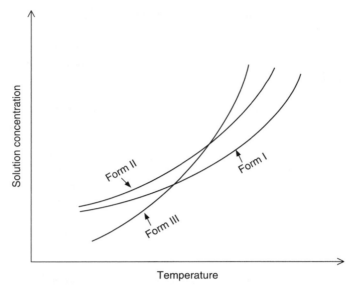

Figure 2-14 Solubility curves of monotropic and enantiotropic polymorphs. For monotropic polymorphs, there is no crossing-over of solubilities of two forms. For enantiotropic polymorphs, there is a crosssing-over.

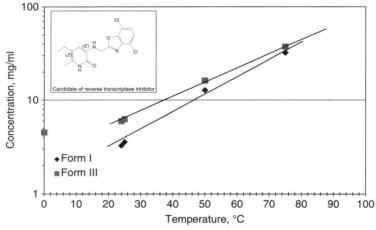

Figure 2-15 Solubility of polymorphs I and III of a reverse transcriptase inhibitor. Below about 80°C, Form III is more soluble than Form I.

shows the microscopic photos of the transformation of Form III crystals to Form I crystals after aging.

As shown in Fig. 2-15, the unstable form, i.e., Form III, could be generated at higher supersaturations. As shown in Fig. 2-16, once generated, the unstable Form III is then converted into a stable form, i.e., Form I.

2.4.2 Measurement and Prediction

Identification of polymorphism is ultimately based upon X-ray powder diffraction pattern analysis. Two forms of crystal must have two different diffraction patterns. Polymorphs

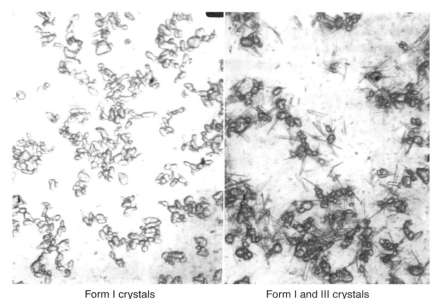

Form I crystals   Form I and III crystals

Figure 2-16   Conversion of a mixture of unstable Form I and Form III crystals to stable Form I crystals after aging.

will also have different melting points, spectroscopic patterns, and differential scanning calorimetric (DSC) profiles (Giron 2000; Clas 2003). Microscopic examination of crystal morphology can be used as well if polymorphs have distinct difference in morphology, as shown in Fig. 2-16. Polymorphs generally have different solubilities, which can also assist in identifying their stability, as shown in Fig. 2-14.

In-situ determination of polymorphism involves process analytical technologies (PAT) (e.g., O'Sullivan and Clennon 2006), as mentioned in Section 2.1.6.

Numerous publications on the prediction of polymorphs appear each year, reflecting the level of interest and importance of this field (Myerson 1999; Vippagunta et al. 2001). Nevertheless, like the prediction of solubility, prediction of polymorphs is challenging and has a long way to go before it becomes a routine practice for industry. Experimental discovery of polymorphs via high-throughput screening devices is the current theme in industry.

2.4.3 Significance for Crystallization

The importance of discovering the most stable polymorphs, if not all polymorphs of the drug at the early stage of drug development goes far beyond crystallization. If a more stable form is discovered in the late phase of drug development, all previous developmental results, including crystallization of pure bulk, need to be reevaluated. For the same reason, if the stable form is discovered after the drug is marketed, the impact can be much more dramatic. Fortunately, the latter scenario is relatively rare (Rouhi 2003).

In the presence of polymorphs, developing a proper crystallization process to control the desired crystal form is a real challenge. A good understanding of solid-liquid equilibrium behaviors under different conditions—for example, the temperature or solvent mixture—is a must. Seeding and control of supersaturation are two critical, if not indispensable, requirements. Example 7-5 shows an example of developing a crystallization process for a polymorphic compound.

As with solid crystallinity, as mentioned in Section 2.3.3, it is not sufficient to grow the desired polymorphs during crystallization. It is necessary to maintain the desired polymorphs throughout the entire operation sequence, including filtration, drying, and milling. For the enantiotrophic system, turning over from a polymorphic form which is stable at room temperature to another polymorphic form which is stable at a higher temperature may occur if the drying temperature is above the crossover temperature. Clearly, it would be desirable to maintain the drying temperature below the crossover temperature.

2.5 SOLVATE

As the name indicates, a compound can incorporate solvent into its crystal lattice and form a different crystal, as in polymorphism. However, the solvate and the original non-solvate material are not identical chemically due to the presence of solvent. Therefore, they are really two different compounds. In practice, almost all solvates are crystalline material, while "anhydrous" material could be either amorphous or crystalline solid.

2.5.1 Phase Diagram

Similar to polymorph materials, nonsolvate materials and solvate exhibit different physical properties and behaviors. Figure 2-17 shows solubility curves of typical nonsolvate and solvate solids as a function of solvent composition and temperature. As shown in the figure, the nonsolvate solid and the solvate solid cross over at composition A_1. Above composition A_1 solvate solid has lower solubility (shown as a solid line) than nonsolvate (shown as a dotted line), and is more stable. Similarly, below composition A_1, a nonsolvate solid has a lower

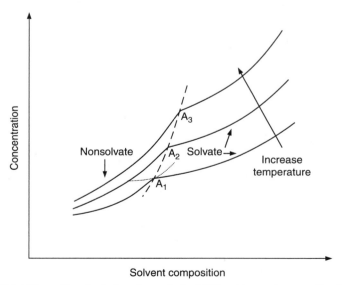

Figure 2-17   Solubility profiles of solvate and nonsolvate at different temperatures as a function of solvent composition. The solvate is more stable at the higher solvent composition, and the nonsolvate is more stable at the lower solvent composition. The crossing-over solvent composition increases as the temperature increases.

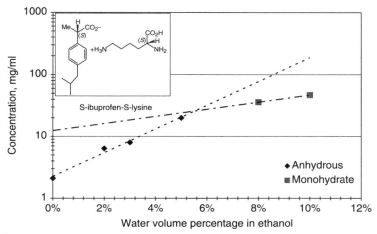

Figure 2-18 Solubility of the anhydrous solid and the monohydrate of ibuprofen lysinate. When the water level is above 5%, the monohydrate is more stable. When the water level is below 5%, the anhydrous solid is more stable.

solubility than a solvate and is more stable. The crossover point A_1 to A_3 forms the crossover line that is a function of temperature. In general, the solvent composition of the crossover point increases as temperature increases.

Solvated crystals are also common in the chemical and pharmaceutical industries. Figure 2-18 shows the room temperature solubility curve of ibuprofen-lysinate as a function of water content in ethanol. As shown in the figure, the crossover point between anhydrous solid and monohydrate is $\sim$5% water. At room temperature, ibuprofen-lysinate remains anhydrous when the water content is below 5% and transforms into monohydrate when the water content is above 5%. In this example, solvate and anhydrous materials also have different crystal habits, as shown in Fig. 2-19.

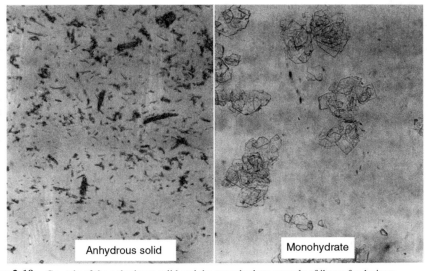

Figure 2-19 Crystals of the anhydrous solid and the monohydrate crystals of ibuprofen lysinate.

A particular solvate is hydrate, in which water molecules form a solid adduct to the parent compound. Hydrate may not be stable below a certain water level, and may lose its water molecule and form anhydrous crystalline or amorphous solids. Above the crossover water level, hydrate is stable and always has lower solubility than the anhydrous solid (Khankari 1995).

2.5.2 Measurement and Prediction

Essentially, all analytical tools mentioned in Section 2.4.2 are applicable for the measurement of nonsolvates and solvates. In addition, the presence of solvent in a solvate can be confirmed by gas chromatography (GC)-mass analysis of the solid.

Special attention should be given in preparing the sample to avoid the solid state transition from solvate to nonsolvate. It is not uncommon to observe that the wet cake solvate loses its solvent during drying. The dry cake becomes a nonsolvate solid, which is not the true solvate solid in the crystallization.

2.5.3 Significance for Crystallization

As in polymorphism, the presence of solvate can present a challenge for drug development beyond crystallization (Variankaval et al. 2007). It is crucial to have a good description of equilibrium solubility behaviors under different operating conditions. Seeding and control of supersaturation are also required to control the crystallization process. Example 7-5 shows the development of the crystallization process in producing the correct polymorph.

As mentioned above, special attention should be given to the drying operation of the wet cake. The wet cake solvate can lose its solvent during drying and turn into a nonsolvate solid that can be either desirable or undesirable. For hydrate, it is generally necessary to maintain the proper humidity in the vapor phase during drying and storage in order to maintain the hydrate form.

2.6 SOLID COMPOUND, SOLID SOLUTION, AND SOLID MIXTURE

When a compound incorporates a second compound other than solvent into its crystal lattice, it creates a new material called a solid compound. As the name implies, solid compounds and solvates are similar phenomena. As with solvates, the ratio of the first compound to the second compound in the solid crystal lattice is fixed. However, in certain cases, the ratio can vary over a wide range. In this scenario, the new compound is called a solid solution. It means that these two compounds can mix freely in the new compound like two miscible liquids. Lastly, a solid mixture is simply a mixture of two compounds physically blended together. Microscopically, each compound still maintains its own crystals. No new compounds or crystal structures are formed.

2.6.1 Phase Diagram

Solid compounds and solid solutions have different physical properties and behaviors from the original compounds. Figure 2-20 is a generic solid–liquid phase diagram in which compounds A and B form an adduct solid compound, C, in the presence of solvent. As shown in

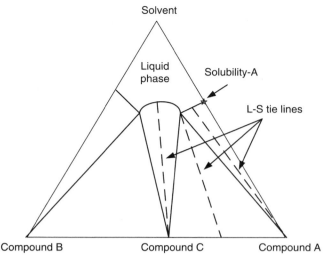

Figure 2-20 The solid–liquid equilibrium relationship of a solid compound. From right to left, the solid phase at equilibrium can be pure compound A, a mixture of compounds C and A, compound C, a mixture of compounds C and B, and pure compound B. The solution phase at equilibrium can be a mixture of compounds A and B at the corresponding liquid–solid tie lines.

the figure, depending upon the overall composition of the system, the equilibrium solid–liquid will have different profiles. On the right portion of the phase diagram, the equilibrium solid will consist of pure compound A and the liquid phase will contain mostly compound A with some compound B (assuming that compound C is dissociated back to compounds A and B in the solution). As the composition moves further to the left, the equilibrium liquid will have a fixed composition of compounds A and B (which is called the eutectic point), but the solid will contain a mixture of compounds A and C. Moving further to the central region, the solid phase will contain only compound C, and the liquid will contain both compounds A and B. The description is similar as the system composition moves beyond the central region. The dominant compounds are now is B and C.

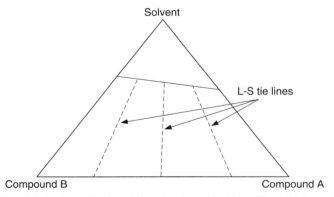

Figure 2-21 The solid–liquid equilibrium relationship of a solid solution. The solid phase at equilibrium will always contains some level of compounds A and B, and so will the corresponding equilibrium solution. In other words, the solubility of individual compound depends directly upon the solid state composition.

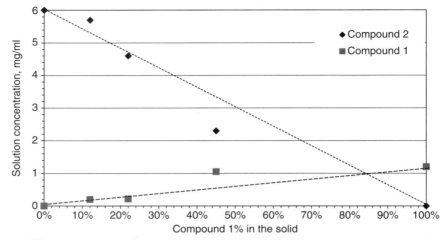

Figure 2-22   Solubilities of chemically similar APIs. This figure shows the dependence of the solubility of lovastatin (compound 1) and simvastatin (compound 2) on the solid composition, which is an ideal example for the solid solution.

Figure 2-21 is a generic solid–liquid phase diagram in which compounds A and B form a solid solution in the presence of solvent. As shown in the figure, the equilibrium liquid composition varies with the equilibrium solid composition. Therefore, as long as the overall composition of the system lies within the triangular area (not along the boundary line), it is impossible to obtain pure solid compounds.

Solid compounds and solid solutions are commonly encountered in optical isomers (Sheldon 1993). Solid compounds and solid solutions can also exist in other structurally similar materials. Figure 2-22 shows the solubility of lovastatin and simvastatin in a solvent mixture of acetone/water. The solubilities of lovastatin and simvastatin vary, depending upon the composition in the solid phase. This is a solid solution similar to the one shown in Fig. 2-21. As we recall from Fig. 2-6, lovastatin and simvastatin differ by only one methyl group. Since lovastatin has lower solubility than simvastatin, the level of lovastatin will be higher in the solid phase, whereas the level of simvastatin will be higher in the liquid phase.

2.6.2   Measurement and Prediction

Solid compounds, solid solutions, and solid mixtures can be differentiated through X-ray diffraction powder analysis and DSC, or other thermoanalytical techniques as mentioned in Sections 2.4.2 and 2.5.2.

Solubility phase diagrams can provide very useful information as well. As shown in Figs. 2-20 and 2-21, the solubility curve can also be used to identify the regions if desirable compounds can be isolated as pure material.

2.6.3   Significance for Crystallization

Solid compounds, solid solutions, and solid mixtures are frequent phenomena for chiral or chemically similar molecules (Wang et al. 2005). From the point of purification via crystallization, the formation of solid compounds or solid solutions presents a real obstacle. Complete rejection of the undesirable compound is impossible, either thermodynamically or kinetically. As illustrated in Fig. 2-22, it is impossible to reject lovastatin as an impurity from simvastatin via crystallization. Therefore, in this scenario, other purification techniques are required.

For the optical isomers, addition of an optically active resolving agent is typically used to "break" solid compounds or solid mixtures. Depending upon the chemical structure of the optical isomer, the resolving agent can be either an acid or a base. The resolving agent can react with the optical isomers (reactive crystallization) and convert the solid compound in its initial neutral form into a solid mixture in its salt form. The resulting solid mixture can then be separated via the solubility difference (Sheldon 1993). Alternatively, if the equilibrium solubility difference between the two salts is insufficient to provide a reasonable yield, preferential crystallization can be performed to separate the diastereomer salts kinetically. Example 7-4 presents shows the resolution of ibuprofen-lysinate diastereomer salts.

2.7 INCLUSION AND OCCLUSION

Inclusion or occlusion takes place when impurities or solvents are physically trapped within crystals. This situation is different from the case of solid compounds. In inclusion or occlusion, impurities can be occluded sporadically within crystals, whereas in solid compounds, impurities are distributed throughout the crystal lattice. While both inclusion and occlusion enclose solvent or impurities in the crystals, inclusion is more applicable to cases where solvents or impurities are trapped within crystal cavities during crystallization, whereas occlusion is more applicable to cases where surface liquid is trapped within crystal clusters or agglomerates during drying. In this book, the terms inclusion and occlusion will be used interchangeably.

2.7.1 Mechanism

Inclusion can occur as a result of rapid crystal growth under high supersaturation (Mullin 2001, pp. 284–286). It is speculated that under rapid crystal growth conditions, a crystal will form multiple growing layers at different heights on the crystal surface. The higher-level layers may grow faster laterally along the crystal surface than the lower-level layers. As a result, the outgrowing of higher-level layers can seal off the surface of crystals and leave cavities within the crystal lattice.

Inclusion can also occur in crystal agglomerates during crystallization. Again, it is hypothesized that under rapid crystal growth conditions, particle-particle contact in the crystallizer slurry results in agglomerates of the original primary particles, and solvent can be trapped within the agglomerates.

Figure 2-23 shows a microscopic photo of a COX-II inhibitor drug candidate. These crystals are generated in the laboratory by rapidly mixing the batch with the antisolvent. As shown in this figure, these crystals form internal cavities which trap the solvents from the crystallization. Similarly, occlusion can occur as a result of crystal agglomeration during crystallization. Figure 2-23 also shows a scanning electronic microscopic image of the same compound generated in a different solvent system. As shown in the figure, the primary particles form agglomerates, which trap the solvent within.

It may not be possible to observe the crystal cavities under the microscope. Nevertheless, based upon the authors' experience, the level of residual solvent trapped within the crystal structure can be related to the degree of nucleation over crystal growth i.e., the higher the degree of nucleation, the higher the level of residual solvent. Example

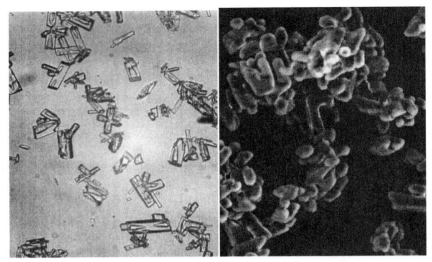

Figure 2-23 Crystal cavity and crystal aggregates formed under unfavorable crystallization and drying conditions.

10-4 is a case study showing that the level of residual solvent can be related to the degree of nucleation over crystal growth.

During drying (of wet cake), occlusion may occur. The presence of occlusion is inferred primarily from the residual solvent profile during drying instead of direct microscopic visualization of occlusion of crystals after drying. Hypothetically, the wet cake can be viewed as a matrix of crystal solids. Both internal pores and the external surface of crystal solids can be covered with a thin layer of solvent. During drying, the external solvent will evaporate and the dissolved solute in the external solvent on the surface can crystallize. If the amount of crystallized material is sufficiently high, it can seal off the external surface of crystal solids. The solvent in the internal space is thus trapped. No further removal of solvent is possible with extended drying. This can also happen with drying of mixed solvent where the antisolvent evaporates before the solvent. It is common to observe agglomerates of dry solid, and the residual solvent in the cake remains steady despite extended drying. To minimize solvent occlusion, it is helpful to keep the drying temperature below certain values so that the amount of dissolved solid in the solvent on the crystal surface is limited. Furthermore, the dryer pressure needs to be maintained sufficiently below the vapor pressure of the solvent at the drying temperature. This ensures effective evaporation of solvent on the surface of wet the cake.

2.7.2 Measurement

HPLC and GC analyses are two commonly used methods for determining the levels of impurities and residual solvent in the cake. Thermal gravimetric analysis is another very powerful tool. It detects not only the level of residual solvent, but also the temperature at which the solvent evaporates. If the cake weight loss due to solvent evaporation occurs at the melting point of the solid, this is a clear indication that solvent is trapped within the cake.

Measurement of impurities in the mother liquor can be done by HPLC. Based upon the overall material balance in the dry cake and mother liquor, it provides valuable information on the trends and distribution of impurities in the cake and mother liquor to differentiate various possible mechanisms during crystallization.

As mentioned earlier, microscopic examination can be used. However, its usage is largely for illustrative purposes.

2.7.3 Significance for Crystallization

From the point of view of purification, occlusion presents a serious obstacle in rejecting the impurities and residual solvent. Since solvent and temperature can drastically affect crystallization behaviors, these variables can play critical roles. Therefore, systematic screening of solvent and identification of proper crystallization conditions for optimum rejection of impurities are desirable. In view of the rapid development of high-throughput screening devices for the measurement of solubility (see Section 2.1.6), it is expected that there will be significant progress in this field in the near future as well.

For solvent systems with a window of operating temperature, proper selection of the method of supersaturation generation (e.g., cooling and antisolvent addition) and mode of crystallization (e.g., batch vs. semicontinuous) can also affect the overall crystal growth rate. In many instances in which solvent or impurity rejection becomes critical, adequate mixing to avoid local high supersaturation can be critical. Examples 9-2 and 10-4 illustrate two cases of rejection of impurities and residual solvent. These examples show how various means are applied to overcome these complications.

2.8 ADSORPTION, HYGROSCOPICITY, AND DELIQUESCENCE

Impurities or solvent can be bound to crystals via a totally different mechanism. The surface of crystals can adsorb impurities or solvents from the surrounding liquid or gas phases. Consequently, impurities or solvent can distribute between the solid and liquid/gas phases. Depending upon the nature of the interaction, various adsorption mechanisms have been proposed (Ruthven 1984).

2.8.1 Phase Diagram

For adsorption of impurities, the surrounding phase of interest is typically the liquid phase. The adsorption isotherm can be built by determining the impurity levels in the solution before and after adding the solid (as adsorbent). Based upon the change of impurity levels in the solution, the adsorption isotherm can then be determined. Figure 2-24 shows a Langmuir-type adsorption isotherm of R-ibuprofen S-lysinate on S-ibuprofen S-lysinate crystals in an ethanol/water solvent mixture. As shown in the figure, a trace amount of impurities can be adsorbed onto the crystal surface even when the impurity concentration in the solution is well below its solubility limit.

For adsorption of solvent, the surrounding phase of interest is typically the vapor phase. Hygroscopicity is a special case of adsorption in which water is preferentially adsorbed onto the crystal surface. Materials that exhibit this behavior are called hygroscopic materials. Deliquescence is a phenomenon in which a hygroscopic material liquefies after adsorbing a certain amount of water onto the crystal surface. Hygroscopicity and deliquescence are commonly encountered in pharmaceuticals.

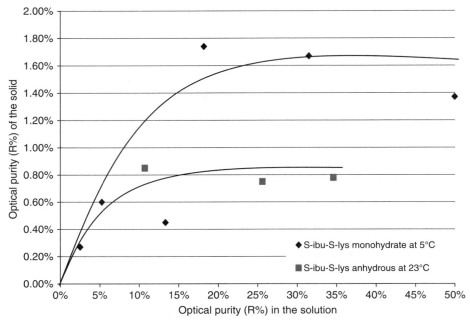

Figure 2-24   Adsorption of R-ibu-S-lys on the crystal surface of S-ibu-S-lys. Adsorption of R-ibu-S-lys on the crystal surface lowers the optical purity of the S-ibu-S-lys solid.

Figure 2-25 shows a typical vapor–solid adsorption isotherm of a hygroscopic material. As shown in the figure, when water concentrations in the vapor and solid phases are located below the isotherm curve, the water in the vapor phase is supersaturated with respect to the water level in the solid phase. Therefore, water in the vapor phase will be adsorbed onto the crystal surface. When water concentrations in the vapor and solid phases are located above the isotherm curve, the vapor phase is undersaturated with respect to the water

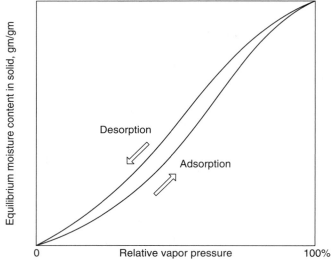

Figure 2-25   Typical vapor–solid adsorption/desorption isotherm.

level in the solid phase. Therefore, water in the solid phase will evaporate into the vapor phase. As a final point, the water vapor pressure, which is in equilibrium with the water adsorbed on a hygroscopic material, should always be less than the water vapor pressure, which is in equilibrium with liquid water.

2.8.2 Measurement

As mentioned in Section 2.7.2, HPLC and GC are two commonly used methods in determining the levels of impurities and of residual solvent in the cake. Thermal gravimetric analysis is another very powerful tool. It detects not only the level of residual solvent, but also the temperature at which the solvent evaporates. If the cake weight loss due to solvent evaporation occurs below the melting point of the solid and near or below the boiling point of the solvent, this is a clear indication that solvent is absorbed on the surface of crystals.

The adsorption isotherm can then be constructed by determining the equilibrium solvent concentration in the vapor and solid phases over a range of solvent composition.

2.8.3 Significance for Crystallization

The efficiency of rejecting impurity via crystallization is reduced if impurities are preferentially absorbed on the crystal surface. For resolution of chiral compounds or crystallization of pure bulk pharmaceuticals, trace levels of adsorbed impurities could be detrimental if not properly treated. Fortunately, surface impurities can generally be removed with proper cake cash. Therefore, an effective wash to remove the adsorbed impurities is highly desirable.

Adsorbed solvent in general will not cause problems in processing since it can be removed during drying. Complications can arise if solvent in the wet cake is occluded within crystals during drying, as mentioned in the previous section. If a solvent mixture is used in crystallization, the interaction of the solvent mixture on the crystal surface is affected by the affinity to crystals and can be quite complicated.

For example, in the drying of a hygroscopic drug compound, the residual solvent profile depends strongly on the relative humidity level in the vapor phase. Figure 2-26 shows the

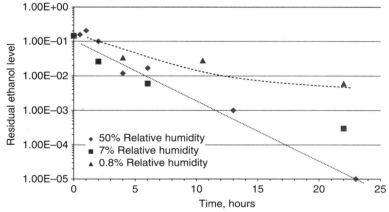

Figure 2-26   Residual ethanol level in the wet cake during drying. At different relative humidities. The drying rate is significantly reduced under low-humidity conditions.

profile of the residual ethanol level in the cake during drying. As shown in the figure, if the relative humidity of the vapor phase is between 7% and 50%, the residual ethanol can be effectively removed from the wet cake. However, if the relative humidity of the vapor phase is below 1%, the residual ethanol in the wet cake will reach a plateau and cannot be removed. If the relative humidity is above 70%, the wet cake liquefies (deliquescence) and loses its crystallinity. This example illustrates the potentially complex interaction of solids and (adsorbed) solvents. It also points out the potential benefit of selective drying of one solvent by controlling the vapor phase composition of the other solvent.

Hygroscopic material requires special attention in processing and handling. In order to remove the water and keep the cake dry, it is essential to maintain a dry environment under all circumstances during drying, packaging, and storage. Otherwise, the cake will pick up moisture from the vapor phase. A wet cake could fail the product specifications or have inferior long-term chemical stability. For certain materials, the cake may deliquesce once exposed to the wet environment and go through an irreversible change from solid state to amorphous material.

The reader may recall that the above scenario the opposite of the case of hydrate, as mentioned in Section 2.5.3. With hydrates, it is desirable to maintain a certain level of relative humidity during drying, packaging, and storage. If the humidity is below a certain level, the hydrate may lose its water and dehydrate.

2.9   CRYSTAL MORPHOLOGY

One of the many unique features of crystals is crystal habits. Crystals can have different shapes and faces that make them unique and distinct from liquid or gaseous materials. In view of the focus of this book, crystal morphology means the approximate shape of crystals. Our purpose is to describe qualitatively the aspect ratios of crystals in three-dimensional space with relevance to the crystallization process. Therefore, we refer the reader to (Mullin 2001, Chapter 1) for a more thorough and comprehensive classifications of crystal morphology.

Crystal morphology refers only to the appearance of crystals. It does not reflect the internal form of crystals, as described in Section 2.4. Crystals of different morphology can have the same or different forms.

2.9.1   General Observations

Roughly speaking, in the variety of pharmaceutical compounds, it is common to observe three types of crystals, as shown in Fig. 2-27.

In this book, the first type crystal is called needle-like because it has only one key dimensional length. The second type of crystal is called plate-like because it has two key dimensional lengths. The third type of crystal is called rod-like or cube-like because it has three key dimensional lengths. Clearly, this is only a qualitative description of the habit of crystals.

From the shape of a crystal, it is possible to infer the surface area and relative growth rates of different crystal surfaces. For needle-like crystals, the surface for crystal growth is primarily at the two ends. The surface along the length of the needle has a much slower growth rate. For plate-like crystals, the surface for crystal growth is at the edges. The surface of the plate has a much slower growth rate. For rod-like or cube-like crystals, all crystal surfaces grow at comparable rates.

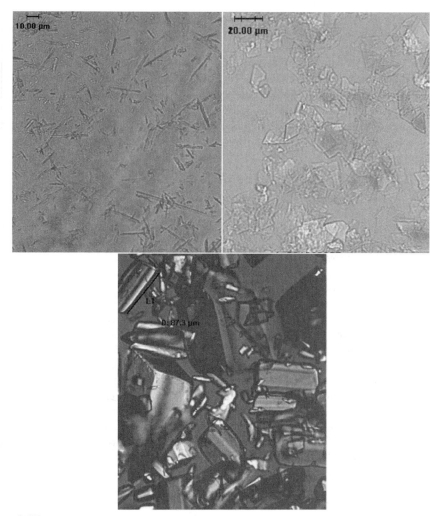

Figure 2-27   Different shape of crystals that may be found during crystallization. They can vary from very thin needle-like crystals to thick cube-like crystals.

The crystal habit gives engineers quick, useful information on the performance of the crystallization process. We will address how to utilize this information in the following sections.

2.9.2   Measurement and Prediction

In practice, optical and scanning electron microscopes are commonly used to view the shapes of crystals. Measurement can be performed offline by taking samples from the crystallizer. Therefore, care should be taken to avoid the complications associated with sampling. Apparatus for in-situ viewing of slurry without sampling is also available commercially. However, the resolution of such devices is limited to tens of microns due to the limitation of the optical focal depth.

Theoretically, it is now possible to predict the morphology of crystals. A number of successful industrial cases have been reported (Myerson 1999). In general, it would require X-ray diffraction data from crystals, which are generated via crystallization experiments. The advances in theory lead to a much better understanding of the internal and external orientation of molecules within the crystal lattice. As a result, cases have been reported in which the shape of crystals can be customized by adjusting the relative growth rates of different surfaces with additives or solvents (Winn 2002). While there is still a long way to go, this is a challenging field and will have a significant impact on industrial practice in the future.

2.9.3 Significance for Crystallization

Understanding the shapes of crystals on the basis of different crystal growth rates on different surfaces is helpful for the development of crystallization processes. Roughly speaking, needle-like crystals have much less surface area for growth. Consequently, more seed is required to provide enough surface area for crystal growth. On the other hand, rod-like or cube-like crystals have much more surface area for growth. So, less seed is required. Furthermore, it is common to observe that the fastest-growing surface outgrows itself and disappears. Therefore, the apparent release of supersaturation by crystal growth is significantly reduced after a certain point of crystallization. If it occurs during crystallization, the supersaturation will build up and a short burst of nucleation may occur. This will lead to a much wider particle size distribution that is generally undesirable.

Needle-like and plate-like crystals create additional process complications. For example, these crystals generally have higher filtration resistance and poorer solid flow characteristics for formulation than cube-like crystals. Therefore, it is highly desirable to grow thicker crystals. To grow "thick" crystals, experimentally, we should try to find the best solvent which favors the formation of such crystals. Meanwhile, solvates and hydrates may form in different solvent environments. Chemical forms, such as salt, free base, and free acid, can also be evaluated. Also, control of release of supersaturation and selection of crystallization conditions to enhance crystal growth over nucleation, which are addressed in the later chapters, would be very helpful.

Another alternative for modifying needle-like crystals is sonication. As shown in Example 7-3, the crystals in the slurry were sonicated repeatedly during multiple crystallization cycles. Sonication breaks one crystal into two or more pieces along its longest dimension. This effectively reduces the length of the crystal or the apparent growth rate in this dimension. Meanwhile, the shorter dimensions of the crystal continue to grow without being affected by sonication. As a result, the crystals grow thicker and shorter over time.

2.10 PARTICLE SIZE DISTRIBUTION AND SURFACE AREA

Particle size distribution (PSD) is another unique feature of crystals. Crystals have different shapes and different sizes. PSD function is a very effective way to describe the distribution of the particle size of crystals over a wide range of sizes. Although crystals have a three-dimensional length, the one-dimensional PSD function is frequently used in practice. Extending the one-dimensional PSD function to two-dimensional or three-dimensional PSD functions may be closer to reality. But these functions are definitely more complex, and their advantages over the one-dimensional PSD function are not always apparent.

Mathematically, the one-dimensional PSD function can be expressed as $n(L)$, where n is the population density function and L is the characteristic crystal length. For cube-like (or spherical) crystals, the characteristic length is approximately the diameter of the crystal. For crystals of other shape there are various definitions for the characteristic length. Most commonly, the characteristic length of the particle with an irregular shape is defined as the equivalent diameter of a sphere which has the same behaviors under the measurement conditions, for example sieving, laser scattering, and sedimentation (Mullin 2001, Chapter 2).

2.10.1 Particle Distribution Definition

By definition of $n(L)$ above, $N(L_1, L_2)$, which represents the total number of particles within the size range between L_1 and L_2, can be expressed as

$$N(L_1, L_2) = \int_{L_1}^{L_2} n(L)dL,$$

while the total concentration of particles, N_T, is given as

$$N_T = \int_0^\infty n(L)dL.$$

All other formulae can be derived based upon the above definition. For example, A, which is the surface area per unit mass of crystal, can be expressed as

$$A = \frac{K_a \int_0^\infty L^2 n(L)dL}{\rho K_v \int_0^\infty L^3 n(L)dL},$$

where K_a is the surface area shape factor, K_v is the volume shape factor, and ρ is the density of the crystal.

For spherical crystals, L is the diameter of the sphere, K_a equals π, and K_v equals $\pi/6$.

For discrete calculations, $n(L)$ can be replaced by $q(x_i)$ and $N(L_i)$ can be replaced by $Q(x_i)$:

$$Q(x_i) = \sum_{j=1}^i q(x_j) \cdot (x_j - x_{j-1})Q(\ln x_i)$$

$$= \sum_{j=1}^i q(\ln x_j) \cdot \ln(x_j/x_{j-1})$$

$$dQ(x_i) = Q(x_i) - Q(x_{i-1})$$
$$dQ(\ln x_i) = Q(\ln x_i) - Q(\ln x_{i-1})$$
$$q(x_i) = dQ(x_i)/(x_j - x_{j-1})$$
$$q(\ln x_i) = dQ(\ln x_i)/\ln(x_j/x_{j-1}).$$

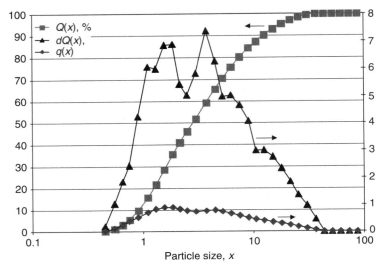

Figure 2-28 PSD functions $Q(x)$, $dQ(x)$, and $q(x)$. This figure shows different way to express PSD: cumulative (Q), differential (q), and incremental cumulative (dQ).

It should be pointed out that $Q(x_i)$ and $Q(\ln(x_i))$, or $q(x_i)$ and $q(\ln(x_i))$, are mathematically equivalent. The former expression is defined over the linear scale of length, whereas the latter expression is defined over the natural log scale of length.

Care should also be taken to differentiate the difference among Q, dQ, and q, as shown in Fig. 2-28.

Very often, the analytical instrument reports only the total number of particles from x_{i-1} to x_i over a range of particle sizes $x_i, i = 1, \ldots, n$. It is a common mistake to treat the data as the function $q(x_i)$ for the subsequent calculation. In reality, the reported value is $dQ(x_i)$ since it represents the total number of particles from x_{i-1} to x_i.

Another common mistake in calculating the PSD data is mixing the linear scale and natural log scale. To calculate $q(\ln(x_i))$, the correct formula is $dQ(\ln x_i)/\ln(x_i/x_{i-1})$, instead of $dQ(\ln x_i)/(x_i - x_{i-1})$.

2.10.2 Measurement

Various methods have been used to measure PSD. These methods include sieving, laser scattering, sedimentation, image analysis, and others. Among these methods, laser scattering devices, such as HELOS or Microtrac (forward scattering device), or FBRM (backscattering devices) are gaining wider use in industry.

Instruments for online measurement of dry and wet systems is also available. Due to the diversity of measurement techniques and the varying definitions of characteristic length, different types of instruments most likely will give different PSDs, for nonspherical particles. Therefore, it should be understood that there is no absolute standard for PSD measurement.

Another problem in PSD measurement is caused by particle agglomerates. A single large agglomerate may contain multiple small crystals. But this agglomerate will be counted as one large particle if it is not dispersed during measurement. Therefore, it is always desirable to perform microscopic examination and surface area measurement in conjunction with PSD measurement.

The surface area of a crystal is typically measured using the Brunauer-Emmett-Teller (BET) technique. The measurement of surface area is much more consistent than that of PSD among various vendors.

2.10.3 Significance for Crystallization

Information on particle size distribution during crystallization can provide valuable insight into the nucleation and crystal growth behaviors of the system. An increase in total particle count and a decrease in particle size could represent nucleation. An increase in particle size with a similar particle count could mean crystal growth. Example 7-2 shows the use of online monitoring PSD for the development of crystallization processes.

Control of PSD is critical for various aspects of drug development. Drugs with low water solubility generally require fine particles with a narrow PSD. Other than milling, special techniques, such as impinging jet crystallization, as shown in Example 9-6, are sometimes employed to meet such requirements. For drugs with high water solubility, different crystal size distributions can increase or decrease the complexity of the formulation process. Therefore, proper control of crystal size by crystallization (and/or milling) is essential.

Chapter 3

Polymorphism

Polymorphism, as applied to the solid state, can be defined as the ability of the same chemical substance to exist in different crystalline structures (Findlay et al. 1951) (regular, repeating arrangement of atoms or molecules in the solid state). The different structures are generally referred to as polymorphs, polymorphic modifications, crystal forms, or forms (Verma and Krishna 1966). Strict adherence to this definition of polymorphism excludes solvates and hydrates (specific water solvate) as polymorphs because they correspond to different chemical substances. Solvates and hydrates are sometimes referred to as pseudopolymorphs. Molecule A is a different chemical substance than molecule A coordinated with a solvent.

3.1 PHASE RULE

This difference is further reinforced by application of the phase rule to the equilibrium between two strictly defined polymorphs of a compound or the equilibrium between a compound and a corresponding solvate of that compound. In the former case, there is only one component (in the phase rule sense—the compound). There are two phases (the two polymorphs) and, therefore, there is only one degree of freedom for equilibrium between two polymorphs by application of the phase rule equation

$$F = C - P + 2 \tag{3-1}$$

where F is the degrees of freedom of the system (number of variables that must be specified to define the system at equilibrium, such as temperature, pressure, concentrations, etc.), C is the number of components, and P is the number of phases. In the solvate case there are two components (the compound and the solvent), and there are, again, two phases (the compound and the solvate of the compound); therefore, there are two degrees of freedom in this situation. In characterizing the equilibrium of a compound and a solvate of that compound, more information is required to define the equilibrium situation than for the case of equilibrium between two strictly defined polymorphs. For instance, both the temperature and composition of the solvate (mono, disolvate, etc.) are required to specify the equilibrium state between a compound and a solvate of that compound, whereas only the temperature is required to specify the equilibrium state between two polymorphs. However, to exclude solvates from a discussion of crystallization is not realistic. But the distinction between polymorphism and solvation is clear and has implications, both practical and theoretical, that must be considered.

Crystallization of Organic Compounds: An Industrial Perspective. By H.-H. Tung, E. L. Paul, M. Midler, and J. A. McCauley
Copyright © 2009 John Wiley & Sons, Inc.

3.2   PHASE TRANSITION

3.2.1   Enantiotropy and Monotropy

There are two types of polymorphism from the thermodynamic viewpoint: enantiotropy and monotropy. With monotropic forms, one polymorph is the stable solid modification over the entire temperature range up to the melting point; therefore, there is no real transition point below the melting point region. A virtual or extrapolated transition point above the melting point results from thermodynamic consideration of monotropic polymorphs. With enantiotropic crystal forms, there is a reversible transition point (equilibrium point) at some temperature below the melting point of either polymorph; therefore, both polymorphs have a definite range of temperature over which they are in the thermodynamically stable solid phase below the melting point of either polymorph. Parenthetically, it should be pointed out that when the word stability is used in a discussion on polymorphism, it refers to thermodynamic stability (Gibbs free energy) and not to reactive or degradative stability.

3.2.2   Metastable Equilibrium and Suspended Transformation

In the real world, the transition or equilibrium point between two enantiotropic polymorphs is often not directly observed because of suspended transformations in connection with metastable equilibrium. Measurement of various physical properties at different temperatures can be related to the relative free energy of the polymorphs, and the forms, can be classified as enantiotropic or monotropic.

The concepts of metastable equilibrium and suspended transformations are extremely important in any practical understanding of polymorphism. Metastable equilibrium can be defined as a state which will exist for some time period without change, even though a more stable state does exist (Findlay et al. 1951; Zernike 1955). This is distinct from unstable equilibrium that results in spontaneous and instantaneous change. A suspended transformation is one that should occur on the basis of the thermodynamic considerations but does not occur because of kinetic factors. Suspended transformations must and do result in metastable states. Without metastable equilibrium and suspended transformations, polymorphism would not be as significant an issue as it is.

Closely allied with the concepts of metastable equilibrium and suspended transformations is Ostwald's rule (Ostwald's step rule or "law" of successive reactions). Essentially Ostwald's rule states that in all processes it is not the most stable state with the least amount of free energy that is initially obtained but the least stable state lying nearest to the original state in free energy (Ostwald 1897). It is easy to see how this rule and the concept of suspended transformations can explain the production of a metastable polymorph through crystallization from a melt or solution.

3.2.3   Measurement

Detection and characterization of polymorphs and/or solvates rely on various experimental techniques. X-ray powder diffraction (XRPD), solid state nuclear magnetic resonance (NMR), solid state infrared (IR) and solid state Raman are useful in demonstrating differences in the solid state. Thermal analytical techniques, including differential thermal analysis (DTA), differential scanning calorimetry (DSC), and thermogravimetry (TG), are also useful in indicating differences in the solid state and can be helpful in distinguishing polymorphs from solvates and hydrate. Gas chromatography (GC) and element analysis are also helpful in differentiating polymorphs and solvates. Solution calorimetry and microscopy, both electronic and optical, have also been extensively used. There are many other experimental techniques that have been employed in the study of polymorphism and solvates. A complete list of these techniques is almost impossible. The choice of experimental techniques can be highly individualized and dependent on the investigators' scientific background. Some techniques, however, are almost universally used by experimenters because of their convenience and widespread availability.

DTA and DSC are particularly useful in determining the relative stability of polymorphs in the temperature region of the melting points of the polymorphs. The higher-melting polymorph is clearly the more thermodynamically stable form in the temperature region of the melting points. In the absence of an observable endothermic transition with increasing temperature between polymorphs with DTA or DSC, it is not possible to ascertain from the melting points alone the relative stability of the polymorphs in the lower temperature range (i.e., room temperature). Measurement of the heats of fusion (ΔH_f) of the polymorphs with DSC or quantitative DTA or other suitable methods can give a good indication of which polymorph is most likely to be more stable at lower temperatures. The polymorph with the higher ΔH_f is most likely the more stable form at lower temperatures. If the higher-melting polymorph also has the higher ΔH_f, then that polymorph is almost certainly the most stable one at all lower temperatures. Then under these conditions, the polymorphs would be related as monotropic polymorphs, excluding any extremely nonregular thermodynamic behavior. If the lower-melting polymorph has the higher ΔH_f, then there will be a transition point between the two forms at some lower temperature. That makes these forms enantiotropic polymorphs, again excluding any extremely nonregular thermodynamic behavior. This type of reasoning has been formularized into the heat of fusion rule, which was proposed by Burger and Ramberger (1979).

The following equation can be use to estimate the transition temperature for two polymorphs:

$$T_t = (\Delta H_{f\,\mathrm{I}} - \Delta H_{f\,\mathrm{II}})/(\Delta H_{f\,\mathrm{I}}/T_{\mathrm{I}}^{\mathrm{o}} - \Delta H_{f\,\mathrm{II}}/T_{\mathrm{II}}^{\mathrm{o}}) \qquad (3\text{-}2)$$

where the T^{o}s are the melting points of the forms in absolute or Kelvin degrees and the subscripts I and II refer to the two polymorphs. If the resulting temperature is higher than either melting point, the polymorphs are monotropic. If the calculated temperature is lower, then the polymorphs are enantiotropic. Of course, Equation 3-2 involves some assumptions, such as the heat of fusion being constant with temperature. A more extensive formulation for inferring the stability relationship of polymorphs has been proposed (Yu 1995) that attempts to eliminate certain assumptions used in developing Equation 3-2.

An alternative method of ascertaining the relative stability of polymorphs at lower temperatures, which can also provide some practical data that could be used in designing efficient crystallization processes, is measurement of the solubility of the forms as a function of temperature under nonconverting conditions. The following chemical equations, along with the corresponding free energy relationships, illustrate the reasoning behind the solubility approach.

$$\text{Polymorph Solid (I)} \iff \text{Solution } (a_{\mathrm{I}}) \qquad (3\text{-}3)$$

$$\text{Solution } (a_{\mathrm{I}}) \longrightarrow \text{Solution } (a_{\mathrm{II}}) \qquad (3\text{-}4)$$

$$\text{Solution } (a_{\mathrm{II}}) \iff \text{Polymorph Solid (II)} \qquad (3\text{-}5)$$

$$\text{Polymorph Solid (I)} \longrightarrow \text{Polymorph Solid (II)} \qquad (3\text{-}6)$$

Equation or reaction 3-3 corresponds to solid polymorph I in equilibrium with its saturated solution with a corresponding solute activity of a_I. The free energy change (ΔG) for this reaction is zero because it is at equilibrium. Reaction 3-4 is a concentration or dilution of the solution of the compound with an activity of a_I to a solution with an activity of a_{II}. The ΔG for this reaction is related to the activity ratio and is equal to $RT \ln(a_{II}/a_I)$. The next reaction is another equilibrium reaction, but this time between solid polymorph II and its saturated solution of activity a_{II}. The free energy change is again zero. The sum of these three reactions yields the reaction of solid polymorph I to solid polymorph II. The free energy change for the conversion is the sum of the ΔGs for the three preceding reactions and is equal to $RT \ln(a_{II}/a_I)$. If the solutions are assumed to be ideal, then the relative activities (a_{II}/a_I) can be equated with the relative solubilities (S_{II}/S_I). If S_{II}/S_I is greater than 1, then polymorph II is less stable than polymorph I. If the solubility ratio is less than 1, then polymorph II is more stable than polymorph I. The solubility determinations must be made under conditions where no conversion of the solid phase occurs. Experimentally, this requires that the solid phase, after a suitable equilibration period with the saturated solution, be examined experimentally by a method that can distinguish the polymorphs, such as XRPD, solid state NMR, etc. Simultaneous measurement of the concentration of the compound in the saturated solution provides the solubility under the conditions of the experiment. The sublimation pressure of the polymorphs at any given temperature can also be used to decide which polymorph is more stable. However, sublimation pressures of solids are usually low and are more difficult to measure experimentally than solubility.

3.3 EXAMPLES

One of the most celebrated case of polymorphism is that of graphite and diamond. Both are crystalline modifications of elemental carbon. Graphite is thermodynamically more stable than diamond at room temperature. The free energy difference is small, amounting to only 0.69 kcal/mole at 25°C and 1 atmosphere (Angus and Hayman 1988). This free energy difference is only slightly higher (~16%) than the corresponding RT value (0.59 kcal/mole). It is obvious that there is a suspended transformation involved and that that implies that there is a relatively large activation energy (kinetic) barrier preventing the conversion of diamond into graphite at room temperature. This case illustrates the importance of metastable equilibrium and suspended transformations. It also demonstrates that the physical properties of the polymorphs can be different and subjected to measurement even though one polymorph is metastable with respect to the other. Diamond is certainly harder and has higher thermal conductivity at room temperature than graphite.

Of course, diamond and graphite are not the only examples of polymorphs. In fact, polymorphism is quite prevalent among pharmaceutical compounds. Solvation and hydration also frequently occur with medicinally active substances (Verma and Krishna 1966; Byrn 1982; Angus and Hayman 1988). In fact, it is rare when a medicinal agent exhibits only a single crystalline structure (Ostwald 1897; Haleblian and McCrone 1969; Ip et al. 1986; McCauley 1991). The different solid forms of the medicinal agent will have different physical properties and as a consequence may have different bioavailabilities. Therefore, control of the solid form of the medicinal agent is extremely important to provide consistent biological activity (Yang and Guillpry 1972). To ensure, as far as possible, an unchanged bulk material and final pharmaceutical product upon storage and for shelf life, it is wise to choose the most stable polymorph or solid phase under the conditions of storage, usually near room temperature. Examples of polymorphism among pharmaceutical compounds follow. This, of course, is not an all-inclusive list and in fact only skims the surface of examples. These cases, however, will help to illustrate some of the concepts developed above and indicate the experimental data required to classify the polymorphs and their relative stability. They will also address some issues that may affect the isolation and production of the desire solid phase.

> **EXAMPLE 3-1**   *Indomethacin*

PROPERTIES AND CHARACTERIZATION

Indomethacin (Shen et al. 1963), one of the first nonsteroidal anti-inflammatory drugs (NSAIDs), is known to exist in three different solid forms. The three forms have different XRPD patterns, and additional chemical analysis indicates that there are no solvates involved with these three forms. Different DTA melting point behaviors (Figs. 3-1 and 3-2) are also observed, as predicted. Forms I and II show single melting point endotherms with peak temperatures of 163°C and 157°C, respectively. Form III shows an initial melting endotherm with a peak temperature of 151°C followed by an exotherm that corresponds to the crystallization of Form I, which then melts, yielding the endotherm with a peak temperature of 163°C.

The classic DTA technique permitted visual observation of the sample during the experiment; thus, apparent melting and recrystallization were readily detected. Additional experiments, of course, confirmed that the exotherm was the recrystallization to Form I. The heating of a sample of Form III was stopped after the initial melting and recrystallization of the sample. The resulting solid sample was found to be Form I by XRPD and solid state IR as anticipated from the original DTA curve. From the DTA data, it is clear that Form I of indomethacin is the most stable form in the region of the three observed melting points.

Solubility data at 25°C (Table 3-1) indicate that Form I is also the most stable form at the lower temperatures. The solubility measurements were performed in water at constant temperature and pH because indomethacin is a carboxylic acid and changes in pH would affect its solubility, as would changes in temperature.

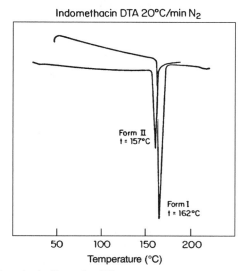

Indomethacin DTA 20°C/min N$_2$

Form II
t = 157°C

Form I
t = 162°C

Temperature (°C)

Figure 3-1   DTA for indomethacin (Forms I and II).

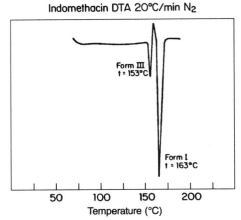

Figure 3-2   DTA for indomethacin (Form III).

Table 3-1   Solubility (mg/ml) of Indomethacin Forms I, II, and III in Aqueous Phosphate Buffer at 25°C

pH	Form I	Form II	Form III
5.6	0.031	0.049	0.073
6.2	0.109	0.159	0.226
7.0	0.545	0.814	1.15

From these results, it was concluded that the pairs of polymorphs are monotropic and that Form I is the most stable modification of the three indomethacin polymorphs.

ISOLATION AND CRYSTALLIZATION

Isolation of the individual polymorphs of indomethacin was initially accomplished by trial and error. There was no absolute way of predicting that there would be polymorphism with indomethacin or any other compound, for that matter. Initially, crystallization under different conditions for process reasons, not for the search for polymorphs, produced different solid phases that were shown to be polymorphs of indomethacin, as indicated above.

Generally, Form I results from controlled crystallizations, either by slow cooling with adequate mixing or slow addition of an antisolvent with adequate mixing at a constant temperature. Form II usually results from rapid crystallizations with either temperature or antisolvent. Even after initial isolation of Form III of indomethacin, it was difficult to predict when it would be obtained, but the conditions were usually somewhat chaotic. Serendipity appears to be a major factor in the isolation of Form III.

Once the indomethacin forms are available, seeding a crystallization batch with the desired form will usually result in the isolation of that form. When excess solid Form II and Form III, either individually or as a mixture, are stirred (agitated, equilibrated) in most solvents and for sufficient time, they convert to the more stable Form I. The time to conversion is roughly related to solubility; conversion time is faster for the solvents with higher indomethacin solubility. In water (Table 3-1), where the solubility is relatively low, Form II and Form III of indomethacin did not convert even after 24 hours of stirring.

EXAMPLE 3-2 *Sulindac*

PROPERTIES AND CHARACTERIZATION

Sulindac (Shen et al. 1971) is another NSAID and exists in two nonsolvated solid modifications. The chemical consistency of the two solid phases was demonstrated by various methods, including solution NMR. The two polymorphs were designated Forms I and II of sulindac. Differences in the solid state were detected by XRPD and DTA (Fig. 3-3).

It is clear from the DTA curves that Form I is the more stable polymorph in the high-temperature range. However, based upon solubility and rate of dissolution measurements (Tables 3-2 and 3-3 and Fig. 3-4), the reverse situation applies at lower temperatures. Form II is more stable near room temperature. The extrapolation of the solubility and dissolution rate date indicate a transition point at about 165°C for Form II to Form I. This transition is not observed in the DTA curve for Form II, as it should be if the situation were thermodynamically ideal. The lack of transition in the DTA curve corresponds to a case of suspended transformation. Forms I and II of sulindac are enantiotropic polymorphs, with Form II being more stable at room temperature (McCauley 1991).

ISOLATION AND CRYSTALLIZATION

Controlled crystallization from solutions of sulindac in the lower temperature region, below the transition temperature (actually, less than 100°C because no solvent crystallizations are performed at higher temperatures), ordinarily results in the formation of Form II of sulindac.

Rapid crystallization in the lower temperature range tends to yield Form I of sulindac. The use of chlorinated solvents (methylene chloride, chloroform, 1,1-dichloroethane, etc.), either under carefully controlled conditions or more chaotic circumstances, results in the isolation of Form I of sulindac after drying off of the solvent. The initially isolated solid in these cases was determined to be a sulindac solvate of the corresponding chlorinated solvent. Once the solvates are broken by drying, Form I of sulindac results, indicating a strong structural relationship between Form I and the solvates.

In nonsolvating solvents at room temperature and at higher and lower temperatures, as long as the temperature is below the apparent transition temperature ($\sim$165°C), it is possible to convert Form I of sulindac to the more stable Form II through extended agitation (stirring, equilibration), similar to the indomethacin case. This conversion to the more stable form with extended equilibration in nonsolvating solvents is, in general, true for most metastable polymorphs. The time to conversion is usually inversely related to the solubility.

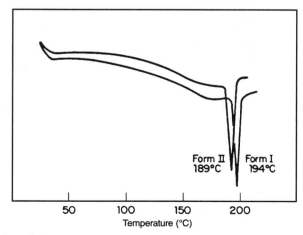

Figure 3-3 DTA for sulindac.

Table 3-2   Solubility of Sulindac Forms I and II in Ethyl Acetate

Temperature (°C)	Form I (mg/ml)	Form II (mg/ml)
6	5.43	2.97
25	6.76	4.02
44	10.40	7.10
54.5	13.85	9.38

Table 3-3   Dissolution of Sulindac Forms I and II in 29.3% *n*-Propanol – Water

Temperature (°C)	Form I (mg/min)	Form II (mg/min)
16.2	0.2069	0.1253
	0.4678	0.3112
35.0	1.1369	0.7732
	1.6860	1.1369

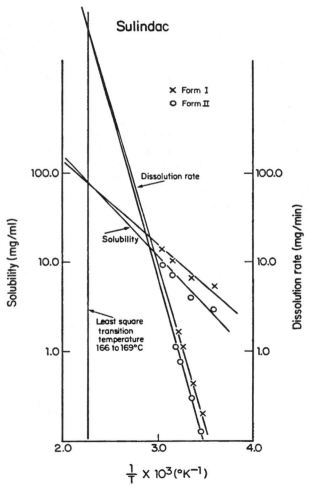

Figure 3-4   Solubility and dissolution rate for sulindac.

EXAMPLE 3-3   *Losartan*

BACKGROUND

Losartan, a potassium salt and an orally active antihypertensive agent, exhibits polymorphism with two known forms (Raghavan et al. 1993). When losartan is crystallized from solution (usually a mixture of cyclohexane, isopropanol, and water), only one anhydrous form (Form I) is observed. Variations in the solvents, temperature (~80°C and lower), and rate of crystallization appear to have no effect on the nature of the solid phase.

PROPERTIES AND CHARACTERIZATION

When the solvent-isolated losartan was subjected to DSC characterization, the DSC curve (Fig. 3-5, curve A) showed a minor endotherm at an extrapolation onset temperature of about 229°C (10°C/min) and a major melting endotherm at an extrapolation onset temperature of about 273°C (10°C/min). When a sample of Form I was heated to 255°C and then cooled back to room temperature, the subsequent DSC curve showed only the high-temperature endotherm (Fig. 3-5, curve B). Chemical analysis (HPLC) and solution NMR showed no change in the material heated to 255°C and cooled back to room temperature. However, XRPD indicated a change in the crystal structure. Therefore, it was concluded that the minor endotherm corresponded to a kinetically irreversible enantiotropic polymorphic transition and that the losartan system was not under complete thermodynamic control. Form I is the low-temperature stable form, up to the transition point, and the high-temperature stable form was

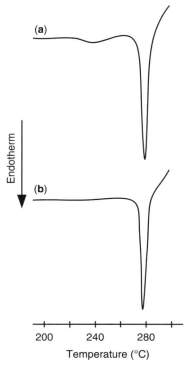

Figure 3-5   DSC thermograms for losartan (a) before and (b) after heat treatment at 255°C. The heating rate was 10°C/min. The material before heat treatment is identified as Form I and the heat-treated material as Form II in the text.

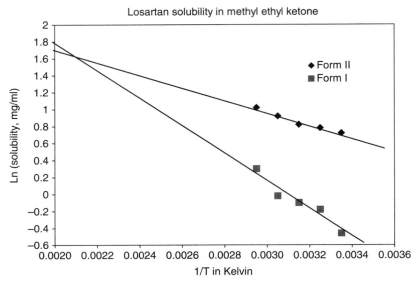

Figure 3-6   Solubillity data for losartan.

called Form II of losartan. DSC curves of Form I at various slower heating rates showed a decrease in the extrapolation onset of temperature of the minor endotherm with no significant effect on the position of the melting endotherm, again indicating kinetic control of the minor endotherm.

The solubility measurements of the two losartan forms (Form II prepared by heating) in isopropanol and ethyl acetate at 25°C and with overnight equilibration resulted in the conversion of Form II to Form I and no change in the initial Form I. The observed solubilities were ~35 to ~0.3 mg/g, respectively. However, in isopropyl acetate, no conversion of either form was observed after overnight equilibration at 25°C. The resulting solubilities were 18 and 41 mcg/gm for Forms I and II, respectively. More extensive solubility data, in methyl ethyl ketone at temperatures ranging from 25 to 65°C with shorter equilibration times that did not result in any conversions, showed Form I, once again, to have the lower measured values (Crocker and McCauley 1997). Extrapolation (Fig. 3-6) of the methyl ethyl ketone data indicated that the two solubility-temperature curves cross at a temperature of about 192°C, which is consistent with the enantiotropic transition seen in the DSC curves for Form I. The solubility data confirm that Form I is the more thermodynamically stable form at room temperature.

Comprehensive spectral analysis including solid state FTIR, solid state Raman, and solid state ^{13}C NMR by Raghavan et al. (1993), resulted in the following conclusion, particularly from solid state ^{13}C NMR: "... spectral characteristics of Form I were interpreted in terms of the presence of more than one orientation for the *n*-butyl side chain and the imidazole ring. In addition, the spectral characteristics of Form II were consistent with a large molecular motion of the *n*-butyl side chain." Although spectral differences were observed, no conclusions about relative thermodynamic stability could be or were made from the spectral data, leaving those conclusions to the more traditional methods of thermal analysis (DSC) and solubility measurements.

EXAMPLE 3-4   *Finasteride*

PROPERTIES AND CHARACTERIZATION

Finasteride (Rasmusson et al. 1986), a benign prostatic hypertrophy agent, also exists in two polymorphic modifications, designated Forms I and II of finasteride. Chemical consistency was assured by various ancillary analytical methods. Solid state differences were demonstrated by XRPD (Fig. 3-7) and DSC (Fig. 3-8) (McCauley 1991).

The DSC curve for Form II shows a single melting point endotherm with an extrapolated onset temperature (T_{ons}) of 257°C and an associated heat of fusion of 88 J/gm. The DSC curve for Form I exhibits a melting point endotherm at an extrapolated onset temperature of 257°C with a heat of fusion of 89 J/gm, which is indistinguishable from the melting endotherm for Form II. Additionally, the Form I DSC curve shows a minor endotherm at about 230°C (at 20°C/min) with an associated heat of about 11 J/gm. XRPD after the minor endotherm and cooling back to room temperature indicates that Form I converted to Form II. There is no evidence, visual or otherwise, of any melting during this minor endothermal process. There are some changes observed in the solid phase, particularly with microscopy, both optical and electronic, in addition to the change in the XRPD pattern. The melting endotherms in both DSC curves correspond to the melting of Form II. A real melting point for Form I of finasteride is not observed.

The DSC curve for Form I of finasteride is characteristic of an enantiotropic form. Form I is more stable at lower temperatures, with conversion to the higher-temperature stable Form II at the transition point. Cooling samples back through the transition point (unmelted) to room temperature and below results in Form II only. Cooling of melted samples, whether starting with Form I or II, results in the crystallization of Form II, with no conversion to Form I upon cooling to room temperature and

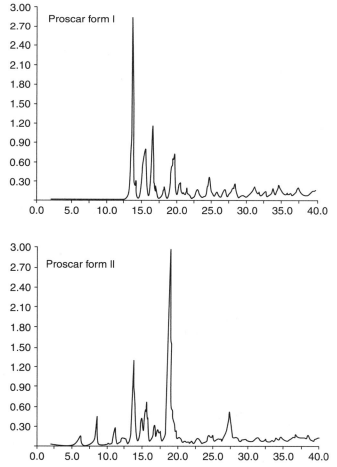

Figure 3-7   XRPD for proscar.

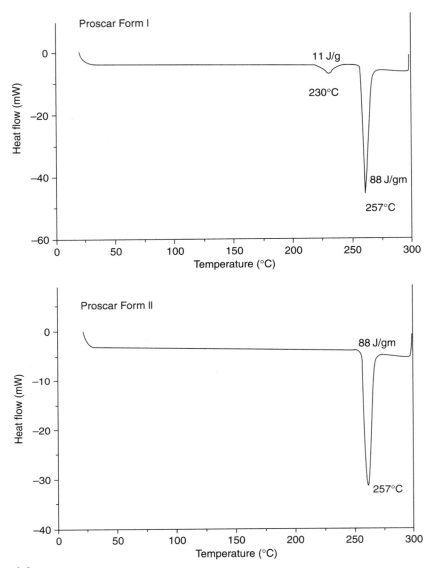

Figure 3-8   DSC for proscar.

below. The lack of transition from Form II to Form I upon cooling is another example of a suspended transformation and indicates that, upon cooling, the finasteride polymorphic system does not exhibit thermodynamically ideal behavior. The relative solubility measurement confirms that Form I is the more stable form at room temperature. The solubilities at 25°C for Forms I and II in cyclohexane are 0.2 and 0.45 mg/gm, and in water they are 0.16 and 0.59 mg/gm, respectively (McCauley 1991).

ISOLATION AND CRYSTALLIZATION

Form I of finasteride can be produced by the addition of water to an acetic acid solution of finasteride such that the final weight % of water is equal to or exceeds 84%. If the weight % of water is less than 83%, acetic acid solvates (mono- and di-) of finasteride are obtained that convert to Form II of finasteride upon drying off of the acetic acid. Form I can also be crystallized (heat-cool) from ethyl or isopropyl acetate containing no water or small amounts of water. The absolute amount of water that can be used varies with the ethyl or isopropyl acetate solvent.

If additional and sufficient water is added to either of the acetate solvent systems, a unique solvate of finasteride containing both water and the corresponding organic solvent is formed, which yields Form II upon drying off of the solvents (Dolling et al. 1999). These di-solvent solvates display a stoichiometry of two finasterides and two waters to one acetate. Spectroscopic information suggests that the acetates in these solvates are not rigidly bound but more likely are trapped in interstitial voids in the crystals. There is no evidence of the existence of any hydrates without the presence the ethyl or isopropyl acetate molecules. Therefore, the phase diagram for this type of behavior (di-solvent solvate) requires that a unique solid phase containing both water and the organic solvent be present with the finasteride, whether or not the acetate is tightly bound in the structure.

The single-crystal structures of both Forms I and II are known (Wenslow et al. 2000), but no complete single-crystal data exist for the solvates. Minimized force field calculations based upon the crystal structures indicate a lower total energy for Form I than for Form II of finasteride, which is consistent with the thermodynamic conclusions from melting and solubility data.

EXAMPLE 3-5 *Ibuprofen Lysinate*

PROPERTIES AND CHARACTERIZATION

Anhydrous ibuprofen lysinate, an analgesic agent, provides another example of enantiotropic polymorphs. Figure 3-9 exhibits a representative DSC curve for anhydrous ibuprofen lysinate. The curve has features that are similar to those of the DSC curve for Form I of finasteride. However, the XRPD pattern after heating to a temperature above that of the minor endotherm for anhydrous ibuprofen lysinate and cooling back to room temperature remains unchanged, in contrast to the finasteride

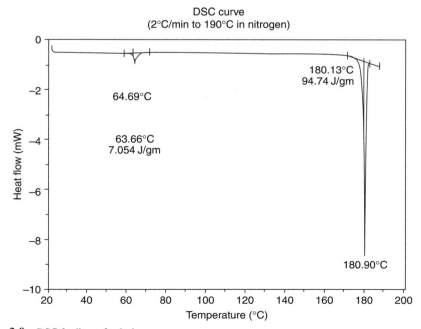

Figure 3-9 DSC for ibuprofen lysinate.

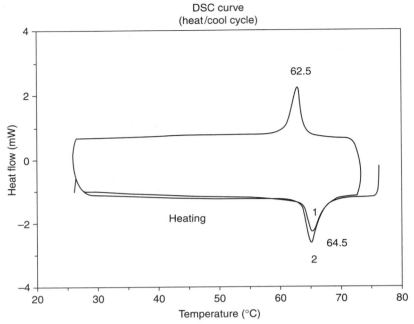

Figure 3-10 Cyclic DSC for ibuprofen lysinate (heat/coal cycle).

Form I case. Using a hot stage XRPD system and obtaining the pattern at a temperature higher than that of the minor endotherm in Fig. 3-9 does show a change in the pattern that converts back to the original pattern when taken again at a temperature below that of the minor endotherm. A cyclic heating and cooling DSC experiment (Fig. 3-10) also shows that the low-temperature endotherm is reversible. There are no suspended transformations for the anhydrous ibuprofen lysinate, either heating or cooling, and the system corresponds to a classical case of enantiotropic polymorphism (McCauley 1991).

ISOLATION AND CRYSTALLIZATION

The stable anhydrous ibuprofen lysinate can be produced by the combination of lysine in water and ibuprofen in ethanol at room temperature such that the final water level in the solvent mixture is below 5 weight %. If the weight % of water is greater than 5%, a monohydrate of ibuprofen lysinate is obtained. The monohydrate converts to anhydrous ibuprofen lysinate if water is flushed out by the anhydrous ethanol.

EXAMPLE 3-6   *HCl Salt of a Drug Candidate*

BACKGROUND

A compound, a class III antiarrhythmic agent, which is a hydrochloride salt and a combination of a substituted heteroatomic tricyclic system with a substituted tetrahydronaphthyl system, provides an interesting example that merges hydration with polymorphism. The compound exists as a solid at room temperature with a melting-decomposition point in excess of 200°C.

When the compound is isolated from certain organic solvents, particularly ordinary alcohols, the corresponding solvate is obtained. When the solvent is removed by drying, the resulting solid is hygroscopic and adsorbs water from the atmosphere to give stoichiometric hydrates. With other organic solvents, notably ordinary ketones and acetates, anhydrous material is isolated directly that is also hygroscopic under certain circumstances. Isolation from water or aqueous-organic systems containing appropriate amounts of water leads directly to various hydrates. The quantity of water required in the aqueous-organic systems is dependent on the particular organic solvent and the temperature of isolation. The water level corresponds to the composition of the eutectic liquid in the three-component phase diagram for the compound, water, and organic solvent at the isolation temperature.

PROPERTIES AND CHARACTERIZATION

Three hydrates of the compound, designated Types A, B, and D, have been found and characterized. The XRPD patterns and the TG curves are represented in Figs. 3-11 and 3-12, respectively.

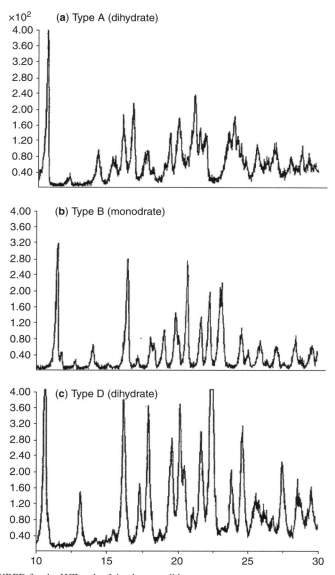

Figure 3-11   XRPD for the HCL salt of the drug candidate.

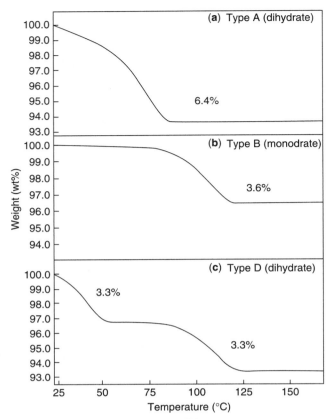

Figure 3-12   TG curves for the HCL salt of the drug candidate.

The TG curve for Type A shows a single stepwise weight loss corresponding to 2 moles of water per mole of compound within experimental error (a dihydrate). Type B exhibits a weight loss corresponding to a monohydrate of the compound. Type D shows two stepwise weight losses, each equivalent to 1 mole of water. Starting with Type D, a hot-stage XRPD pattern of the resulting solid at about 60°C shows conversion to Type B monohydrate. Further heating to about 125°C, on the plateau for the TG curve, yields an anhydrous form of the compound (Type C_B). Type C_B is very hygroscopic and exists only in an essentially anhydrous environment. Exposure to any moisture converts Type C_B to Type B or Type D, depending on the partial pressure of water (relative humidity).

Starting with Type A, the hot-stage XRPD pattern at 125°C is different from that of Type C_B (Fig. 3-13) and corresponds to a second anhydrous form to the compound (Type C_A). Type C_A is stable below 15% relative humidity. Exposure to relative humidity of 15% or above converts Type C_A to the Type A dihydrate. Hot-stage XRPD patterns at temperatures in the region of the observed TG weight loss for Type A show only Type A or mixtures of Type A and Type C_A. There is no evidence of any Type B or any other monohydrate, indicating that Type A converts directly to the anhydrous Type C_A with the loss of two water molecules. This contrasts with the behavior of the Type D hydrate, which loses one water molecule to give the Type B monohydrate, which in turn loses the water molecule to give anhydrous Type C_B.

ISOLATION AND CRYSTALLIZATION

Selecting and producing the solid phase of this compound for pharmaceutical applications indicates how complex these situations can become. The organic solvates, in general, are ruled out based

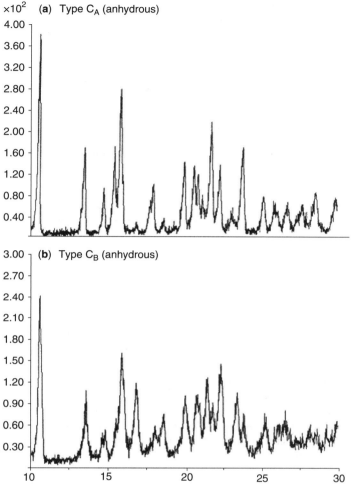

Figure 3-13   Additional XRPD for the HCL salt of the drug candidate.

upon toxicity considerations. Although ethanol solvates have been used, they are normally avoided if other acceptable solid phases are available. Both anhydrous polymorphic modifications, in this case, are too hygroscopic to be used in any ordinary solid pharmaceutical formulation regardless of which one is the more thermodynamically stable phase at room temperate.

Hygroscopicity studies (Table 3-4) with hydrated Types A, B, and D indicate that Type A is the more stable of the hydrated phases over the widest range of relative humidity. Type A will remain unchanged when exposed to the widest range of usual ambient conditions. Clearly, Type A is the more acceptable phase for solid pharmaceutical dosage formulations. However, it is also evident that Types A and D are polymorphic modifications of the dihydrate of the compound. Which one of these polymorphic modifications is the more thermodynamically stable phase at approximately the room temperature was addressed by water solubility measurements in the range of temperature from about 15 to 45°C. The results of the solubility measurements are given in Table 3-5 and indicate that the two dihydrates are enantiotropic polymorphs, with Type D being more stable in contact with a saturated aqueous solution below a transition temperature of about 37°C.

Type D would ordinarily be the prime candidate for isolation by crystallization and pharmaceutical development if it were not for the overdriving considerations of ambient humidity conditions. Type A was chosen for development based upon the humidity situation, but it was clear that because of

Table 3-4 Hygroscopicity Study with the HCl Salt of Example 3-6 at Room Temperature

% RH	Starting with A	Starting with D	Starting with B
10	CA (0.2)	B (3.5)	B (3.4)
20	A (6.3)	B (3.6)	B (3.7)
32	A (6.3)	B (3.7)	B (3.8)
42	A (6.3)	D (6.6)	B (3.7)
58	A (6.3)	D (6.5)	D (6.2)
68	A (6.3)	D (6.6)	
90	A (6.4)	D (7.2)	

Table 3-5 Solubility Study in Water with the HCl Salt of Example 3-6

	Solubility ($\pm$0.2 mg/ml)	
Temp ($\pm$0.1°C)	Type A	Type D
15.2	5.8	4.8
25.2	7.1	6.1
36.1	10.2	9.9
45.7	12.9	14.5

the existence of a more stable dihydrate when in contact with aqueous saturated solutions, it was paramount that the final isolation process be exhaustively studied and sufficiently controlled.

Rapid, spontaneous (no seeding) crystallization from aqueous systems along with immediate filtration, in the temperature range of 20 to 60°C, yielded Type A above or below 37°C, the enantiotropic transition temperature for the two polymorphic dihydrates. Below 20°C, mixtures of Types A and D or just Type D were isolated even with rapid crystallization and filtration. If the crystallization and filtration times are extended at temperatures less than 37°C, once again, mixtures of Types A and D or just Type D are isolated. Type A was reliably obtained by slow, controlled crystallization from appropriate aqueous systems through cooling to and filtering at 50°C. Seeding with Type A under these conditions is an added precaution (McCauley et al. 1993).

EXAMPLE 3-7   *Second HCl Salt of a Drug Candidate*

PROPERTIES AND CHARACTERIZATION

The polymorphism encountered with 1-(5-{[(2R,3S)-2-({(1R)-1-[3,5-Bis(trifluoromethyl)phenyl]ethyl}oxy)-3-(4-fluorophenyl)morpholin-4-yl]methyl}-2H-1,2,3-triazol-4-yl)-*N*,*N*-dimethylmethanamine hydrochloride (Compound A, hereafter), a high-affinity, orally active h-NK1 receptor antagonist (Harrison et al. 2001), illustates the complexity of solid phase behavior.

Two anhydrous polymorphic forms (designated Forms I and II) of Compound A have been found to exist at approximately room temperature (Wang et al. 2002).

Differences in these two forms are indicated by XRPD (Fig. 3-14) and solid state NMR (Fig. 3-15). The DSC curves (Fig. 3-16) also exhibit differences and indicate the existence of additional solid state structural modifications. The curve for Form I shows a minor endotherm at

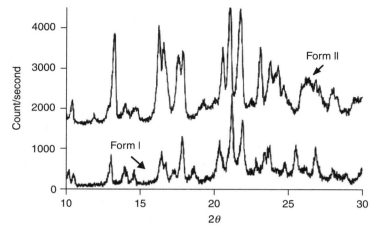

Figure 3-14 XRPD for Compound A, Forms I and II.

about 108°C (T_{ons}; 10°C/min) following by a melting endotherm at about a T_{ons} of 200°C. Further studies have shown that the minor endotherm corresponded to a transition and that the resulting solid phase showed a different XRPD pattern (Fig. 3-17) and was designated Form III. Once formed through heating above 108°C but less than about 130°C, Form III spontaneously converted back to Form I upon cooling down through the temperature region of the initial minor endotherm ($\sim$108°C), indicating that the minor endotherm corresponded to a reversible enantiotropic polymorphic transition. However, once melting occurred, at about 200°C, followed by cooling, no spontaneous recrystallization of any kind was observed.

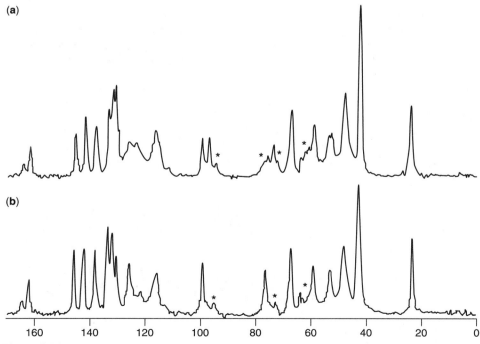

Figure 3-15 Solid-state NMR for Compound A, (a) Form I and (b) Form II (*represents spinning sidebands).

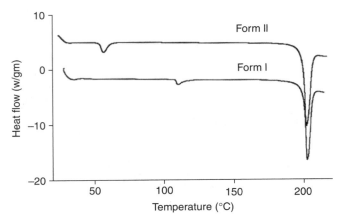

Figure 3-16 DSC scans for Compound A, Forms I and II, at 10°C/min.

The DSC curve for Form II (Fig. 3-16) also showed a minor endotherm, but at a lower T_{ons} (53°C) than seen starting with Form I. The solid phase observed above 53°C was different from any of the previously seen solid structures and was designated Form IV (Fig. 3-17). Form IV, once formed and cooled below 53°C, spontaneously converted to Form II. Forms II and IV were a second pair of enantiotropic polymorphs. The solid state NMR spectra of Forms III and IV were also unique (Fig. 3-18).

Compound A has four polymorphic forms and at least two pairs of enantiotropic polymorphs; Forms I and III are one pair and Forms II and IV are the second pair. The thermodynamic relationship between Forms I and II of Compound A is important and has been ascertained because these forms are the two polymorphs seen in the room temperature region; therefore, they are the ones that would be used in any solid phase formulation. The thermodynamic relationship was established with solubility measurements in tert-butyl acetate and as a function of temperature (Wang et al. 2002). The solubility data are summarized in the plot in Fig. 3-19 and clearly show that Form I is the more thermodynamically stable form in the lower temperature ranges, including room temperature. Extrapolation of the data results in a calculated transition point of about 129°C that is significantly lower than the observed melting point of Compound A and therefore suggest than Forms I and II are a third pair of enantiotropic polymorphs. The lack of any such transition in the DSC curve for Form I is due to the enantiotropic transition to Form III that intervenes at a significantly lower temperature (108°C).

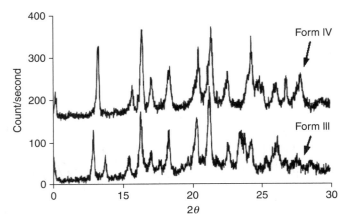

Figure 3-17 XRPD for Compound A, Forms III and IV.

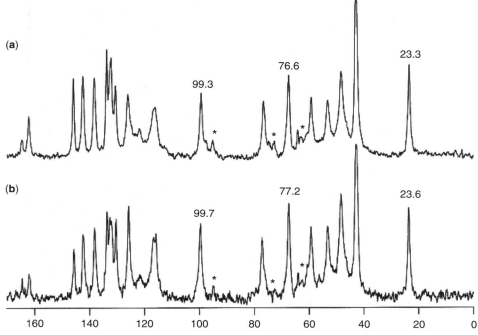

Figure 3-18   Solid-state NMR for Compound A, (a) Form III and (b) Form IV (*represents spinning sidebands).

Establishing the thermodynamic relationship between Forms III and IV of Compound A, although of scientific or academic interest, is not as essential as determining the relationship between the two room temperature forms. In addition, the fact that Form III exists only above $100-110°C$ adds some formidable experimental obstacles to the situation. However, some indication of the relationship can be surmised through a more thorough examination of the DSC curves when starting with Forms I and II (Fig. 3-16). In both cases, the final melting endotherms are essentially identical. Not only are the T_{ons}'s indistinguishable, within experimental error, but the two calculated heats of fusion ($\sim$46 J/gm) are identical, again within experimental error. This suggests that the solid phase involved in the final melting of Compound A is the same whether starting with Form I or Form II. It further implies that at least one more solid phase transition (undetected in the DSC curve) occurs prior to

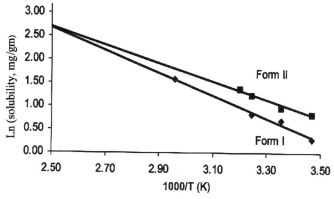

Figure 3-19   Solubility data for Compound A, Forms I and II, in *t*-butylacetate.

melting. In fact, if the solid phase were examined, say with hot-stage XRPD, just before melting (~195°C), one would find that Form IV is present regardless of whether the starting material was Form I or Form II. These findings suggest that Form IV is the final melting form, that Form III has converted to Form IV at some temperature below the melting point of Form IV, and further, that the transition between Form III and Form IV is associated with a relatively low energy or one that is slow enough that it cannot be detected in the DSC curve.

The exact nature of the transition, i.e., enantiotropic or spontaneous monotropic, cannot be completely determined. Again, only some more difficult and demanding experimental data (solubilities or sublimation pressures at high temperatures) would provide unequivocal results. The same, of course, applies to the two remaining polymorphic pairs, Forms I and IV and Forms II and III. Even without knowing the exact nature of all of the polymorphic pairs for Compound A, an interesting experiment can be performed starting with Form I at room temperature and heating it through the Form I–Form III (~108°C) transition and holding at some temperature between 130 and 185°C until Form III converts completely to Form IV, then cooling down of Form IV to room temperature through the Form IV–Form II transition yielding Form II at room temperature, with the resulting implication of the conversion of the more stable form (Form I) to the less stable form (Form II) at room temperature. Of course, there is no real violation of the laws of thermodynamics, only the inaccessibility of Form I to Form II transition (129°C), and the suspended spontaneous room temperature transition confirms that we usually live in a kinetically rather than a thermodynamically controlled world—the real world, not the ideal one.

ISOLATION AND CRYSTALLIZATION

At or near room temperature, controlled crystallization with relatively slow addition of an antisolvent (*n*-heptane) to a solution of Compound A (9/1 ethyl acetate/isopropanol) results in the isolation of Form I. Rapid addition of the antisolvent (uncontrolled crystallization) yields Form II.

EXAMPLE 3-8   *Prednisolone t-Butylacetate*

Prednisolone *t*-butylacetate, a synthetic adrenocortical steroid (Sarett 1956), furnishes an example of the importance of thorough understanding of the final stages of the production process and how they can affect the final bulk solid phase. Isolations of the steroid carried out in anhydrous organic solvents (ethanol, etc.) produced an anhydrous solid form that melted/decomposed above 265°C. The anhydrous prednisolone *t*-butylacetate has very low water solubility, about 40 mcg/ml. Because of this low water solubility, the compound was crystallized out of organic systems containing water in an effort to improve the efficiency (yield) of the isolation process. A monohydrate of prednisolone *t*-butylacetate was isolated when sufficient water was added to the system. Drying of the monohydrate results in an anhydrous form. The anhydrous form converts back to the monohydrate over time at ambient humidities. The original anhydrous form of prednisolone *t*-butylacetate was shown to be non-hygroscopic and could be stirred in water for at least a week without any change. The XRPD pattern (Fig. 3-20) for the original anhydrous form, designated Form III, is clearly different from the pattern for the hydrated material, designated (somewhat inconveniently) Form I. (Some confusion in the concepts of polymorphism and solvation is evident in these unfortunate designations.) On the other hand, the XRPD patterns of Form I and the anhydrous form related to Form I are very similar. Compare the two experimental patterns in Figs. 3-21 and 3-22. The single crystal X-ray structure was available for the monohydrate (Form I). Molecular modeling software was used to predict the XRPD patterns with and without the water. The predicted patterns in both cases are in excellent agreement with the experimental information. The only significant different between the hydrate (Form I) and the related anhydrous material is in the neighborhood of 6 degrees (2θ). Because of the increased yield in the isolation and some pharmaceutical considerations, prednisolone *t*-butylacetate monohydrate (Form I) was selected as the desired solid phase for pharmaceutical formulation.

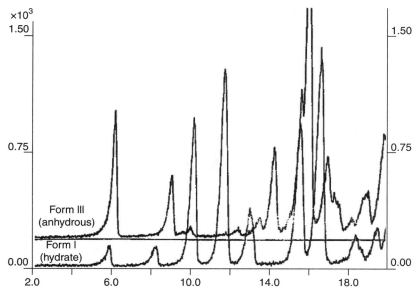

Figure 3-20   XRPD patterns for prednisolone *t*-butylacetate.

Due to the nature of the final formulation and its end use, the prednisolone *t*-butylacetate mono-hydrate had to be recrystallized under sterile conditions. To provide the sterile bulk, the prednisolone *t*-butylacetate monohydrate was dissolved in *N,N*-dimethylacetamide (DMAC), filtered under sterile conditions, and crystallized with sterilized water. Experiments had shown that the solid phase in contact with the saturated solution was indeed the monohydrate. The material was filtered, washed with

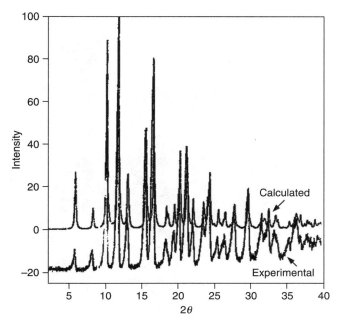

Figure 3-21   Experimental and calculated XRPD patterns for hydrated prednisolone *t*-butylacetate, Form I, with water.

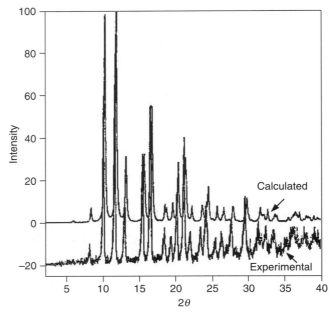

Figure 3-22   Experimental and calculated XRPD patterns for prednisolone *t*-butylacetate without water.

water, and dried, all under sterile conditions. The material was dried until the LOD (loss on drying) of the filtered material minus the water content, as determined by Karl Fischer titration (a standard method for water analysis), reached a specified level. This difference was related to the DMAC content. There was also minimum water content specification to secure the monohydrate. On occasion, the final pharmaceutical formulation, a sterile suspension of the prednisolone *t*-butylacetate monohydrate in

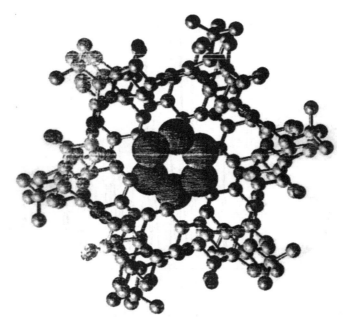

Figure 3-23   Molecular view of hydrated prednisolone *t*-butylacetate.

an aqueous medium, failed to perform as anticipated. The failure involved the aggregation of the suspended solid. The aggregation then resulted in the inability to deliver the suspension through a hypodermic syringe. The failed formulations were found, using XRPD and solid state IR, to contain either Form III (anhydrous prednisolone *t*-butylacetate) or mixtures of Form III and monohydrate. Form III was also detected in the sterilized bulk materials associated with the failed formulations.

An investigation of the washing and drying of the sterile filtered solid with XRPD, solid state IR, and Fourier transform infrared (FTIR)/TG was instituted. The FTIR/TG technique combines the weight loss–temperature profile with analysis of the gases released during the weight loss. XRPD and solid state IR showed that the filtered, washed material was still the monohydrate. FTIR-TG indicated that the DMAC was released from the filtered, washed solid in two distinct temperature ranges. The lower range that corresponded to residual surface DMAC was relatively easy to remove. The second temperature range was significantly higher and appeared to represent DMAC trapped inside the monohydrate crystal structure.

If sufficient DMAC were trapped, it is conceivable that, upon drying at some elevated temperature, the DMAC could preferentially dissolve some of the surrounding prednisolone *t*-butylacetate.

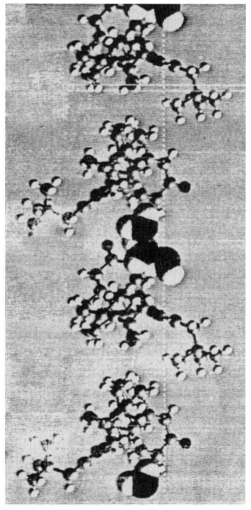

Figure 3-24 Molecular view of one of the crystal faces for hydrated prednisolone *t*-butylacetate.

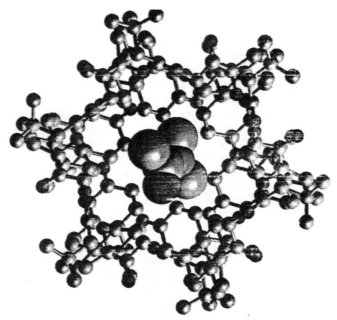

Figure 3-25   Molecular view of prednisolone *t*-butylacetate with DMAC in the water cavity.

When the DMAC was finally evaporated off or the material cooled down, the prednisolone *t*-butylacetate would crystallize to Form III, the undesired anhydrous form.

Examination of the available single-crystal X-ray structure for the monohydrate of prednisolone *t*-butylacetate with molecular modeling software (Molecular Simulations Inc.) showed that the water molecules were present in a cavity (Fig. 3-23). The cavity contained six water molecules, one per one prednisolone *t*-butylacetate molecule. A view of one of the major faces of the crystal (Fig. 3-24) indicated that water molecules were also present at this surface and provided access to the cavity. Using the software, it was possible to show that a DMAC molecule could easily fit inside the cavity (Fig. 3-25), replacing the water molecules. It was reasoned that excess DMAC on the filtered but insufficiently or ineffectively washed material could penetrate into the crystal structure and later cause the crystallization of the undesired From III upon drying. Extensive washing of the filtered material with additional sterile water to improve the effectiveness of this step was suggested, along with a more precise method of determining the DMAC content of the washed material. The latter suggestion was not heeded but the former was and the change led to the desired result: no more failed formulations. The FTIR/TG curve of the extensively washed material indicated a significant decrease in both the loosely held (low-temperature) and the more tightly bound (high-temperature) DMAC.

EXAMPLE 3-9   *Phthalylsulfathiazole*

BACKGROUND

Phthalylsulfathiazole (Moore 1943), an intestinal antibacterial agent, was reported to exhibit polymorphism and/or solvation through the use of thermomicroscopy or hot-stage optical microscopy (Kuhnert-Brandstater 1971). The visual observation reported that upon heating, phthalylsulfathiazole undergoes what appears to be melting followed by recrystallization and then a second melting at a higher temperature. Additionally, it was noted that droplets of liquid were observed on the microscopic cover slide, possibly indicating hydration or solvation.

PROPERTIES AND CHARACTERIZATION

It was also speculated that the initial solid phase was not the most stable phase of phthalylsulfathiazole because of the apparent crystallization to a higher melting phase. In fact, the DTA curve (Fig. 3-26) for phthalylsulfathiazole is similar in appearance to that of Form III of indomethacin (see Fig. 3-3), a known monotropic polymorph of indomethacin. The TG curve (Fig. 3-27) for phthalylsulfathiazole shows a significant weight loss in the temperature region of the first endothermal event seen in the DSC curve. Therefore, this event cannot be a simple melting of the phthalylsulfathiazole. This could, of course, correspond to solvent loss and indicate that the initial phthalylsulfathiazole phase was solvated. However, thin layer chromatography (TLC) and mass spectral analysis showed that the recrystallized material prior to the second DSC endotherm contained no phthalylsulfathiazole and was a mixture of decomposition products. One product corresponded to cyclization with the loss of water to form an imide, and the other resulted from the loss of phthalic anhydride, yielding the free amine (Fig. 3-28).

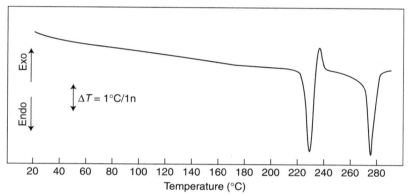

Figure 3-26   DTA for phthalylsulfathiazole.

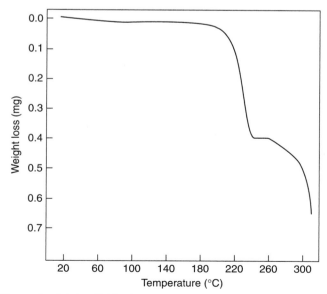

Figure 3-27   TG curve for phthalysulfathiazole.

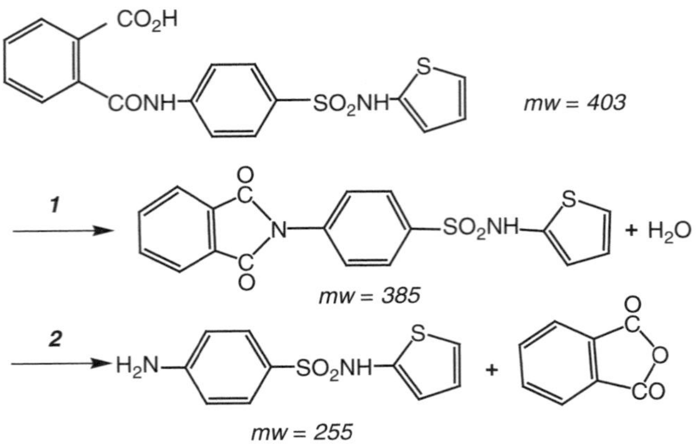

Figure 3-28   Thermal decomposition pathways for phthalylsulfathiazole.

The behavior of phthalylsulfathiazole is a case of decomposition and not polymorphism or solvation. The phthalylsulfathiazole example indicates that reliance on a single experimental technique in the study and classification of polymorphism or solvation is not recommended.

3.4   FUTURE DIRECTION

Increasing awareness of the importance of polymorphism and related issues, particularly in the pharmaceutical arena, has led not only to heightened regulatory surveillance but also to expanded scientific activity. There is extensive activity in the areas of detection, classification, and isolation, and also in the domain of a priori prediction of possible crystal structures and related energy calculations that then attempt to anticipate polymorphism for individual chemical compounds. Although there have been some successes with ab initio methods, they have been associated with less complex and more rigid chemical structures that are increasingly less likely to be encountered in the pharmaceutical industry. However, these methods, with continuing research, will be refined and, therefore, become more useful in the future. Software for the prediction of polymorphism is commercially available from companies such as Accelrys. Accelrys, Inc., was formed (2001) from a combination of other molecular modeling companies including Molecular Simulations Inc., Synopsys Scientific Systems, Oxford Molecular, GCG, and Syomics Ltd.

At the present time, there is no certain way of predicting polymorphism or solvation, and the statement that most medicinal compounds exhibit some form of multiple solid state structures is too general to be useful in a particular situation. The issues involved are of sufficient importance that the discovery of polymorphism in an individual case should not be left, at this stage of scientific advance, to serendipity or happenstance. Isolation and crystallization in a number of solvents with different properties and under a variety of conditions should become an integral part of the initial investigation into the physical properties of any promising medicinally activity compound. Systems are commercially (Symyx Technologies, Inc., and Crystallics B.V.) available or under development that attempt to completely automate the search for polymorphs and solvates.

Chapter 4

Kinetics

As noted in previous chapters, important information is obtained by equilibrium measurements and correlations regarding a given crystallization process. However, much of the effort expended in developing and designing such a process is dedicated to defining the *rate processes* involved. This chapter presents a brief discussion of the rate processes which define the nature of the final crystalline product and, of course, the equipment required to produce it.

Kinetics of nucleation and crystal growth can be modified to some extent to favor an efficient process, but in almost all commercial crystallization processes, this flexibility is limited by the needs of the users of the product.

In the pharmaceutical industry, the users of *bulk final products* are the dosage form formulators, who must provide a product which will produce the desired beneficial effect on patients health. As can be imagined, purity and bioavailability criteria must be satisfied before the formulators can address concerns about other crystal properties such as flowability, compaction properties, and other parameters.

Users of *intermediate products* (those who carry out the next step) are always concerned with levels of impurities being carried forward to them from the previous crystallization process, but generally they have fewer concerns about other properties of the solids. This can often allow manipulation of process parameters to favor reduced time cycles or more efficient types of processing equipment.

As noted throughout this book, the reader is encouraged to review the detailed treatment of kinetics in the many excellent texts on crystallization. Development of many of the kinetic models used by current researchers is described in the texts of Nielsen (1964), Ohara and Reid (1973), Nyvlt et al. (1985), and Söhnel and Garside (1992). More recent summaries and descriptions are those of Myerson (2002), Mullin (2001), and Mersmann (2001).

Kinetic models are also described more briefly by Johnson (2003), who provides an overview of rapid precipitation of organics, and by Davey and Garside (2000), who review all aspects of crystallization in an excellent introductory volume.

4.1 SUPERSATURATION AND RATE PROCESSES

As supersaturation is increased in a solution containing crystalline material or surfaces upon which crystals can grow, growth is the dominant process. However, if/when supersaturation reaches a critical value, new nuclei are rapidly formed, relieve much of the supersaturation and, thereafter, becoming new sites for growth. The control of crystal size in industrial

Crystallization of Organic Compounds: An Industrial Perspective. By H.-H. Tung, E. L. Paul, M. Midler, and J. A. McCauley
Copyright © 2009 John Wiley & Sons, Inc.

operations depends upon the control of nucleation and growth. The mechanisms by which nucleation and growth occur are outlined in later sections of this chapter.

The driving force for crystallization is the excess activity of the solute compared to that in equilibrium with the crystalline solid. The most common supersaturation ratio, for example, is $\gamma_i C_i / \gamma_{is} C_i^*$. For most purposes, it is common to assume an activity coefficient (γ) of 1 and to discuss supersaturation in terms of concentration difference ($C - C^*$) or ratio (C/C^* or $[C - C^*]/C^*$).

The kinetic order in expressions for crystal growth is discussed in Section 4.3.1.6, that for heterogeneous nucleation in Section 4.2.3.1, and that for homogeneous nucleation in Section 4.2.1.

In combination with vessel geometry and mixing as well as the character of the slurry, supersaturation governs the rate and nature of the nucleation and crystal growth processes which will occur. As noted above, this chapter provides a brief discussion of these processes.

Supersaturation, at least locally, is an absolute requirement to drive nucleation and crystal growth. For nucleation to occur, molecular clusters (aggregates) of the dissolved substance must continually form and decompose because of local concentration fluctuations. In a supersaturated solution outside the metastable region, the rate of formation of aggregates exceeds that of their decomposition. In the case of crystal growth, growth takes place when the rate of integration of single molecules or aggregates into an existing lattice structure exceeds the rate of release of these entities from the surface, a process also driven by local concentration fluctuations.

Although supersaturation for crystallization in the pharmaceutical industry is usually measured as a difference or ratio of absolute solution concentrations, the solubility product is most often used for very fast precipitation, which is more common with inorganics.

Crystallization is a combination of nucleation and crystal growth processes. Induction time for nucleation, as measured by current in-line instruments, is the combined time for initial nucleation and subsequent growth to reach instrument detectability. Figure 4-1 shows, for release of a given amount of material, the prototypical shape of the curves of

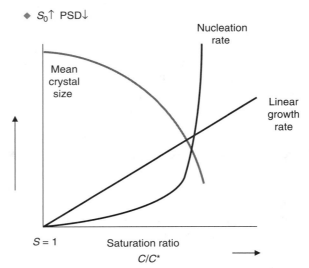

Figure 4-1   Effect of supersaturation on growth rate and particle size.

nucleation and growth rates as they are affected by supersaturation. At low supersaturation, growth dominates and ultimate product particle size is maximized. An increase in supersaturation, shown in Fig. 4-1 as the saturation ratio ($S = C/C^*$) rather than [$(C - C^*)/C^* = (S - 1)$], increases nucleation rates, and at high supersaturation the process is controlled by nucleation mechanisms.

In almost all industrial crystallizations, both nucleation and growth contribute to the final result (purity, morphology, size distribution) in a major way. All growth processes, emphasized in this book as being desirable for maximum control and robustness in many or most situations, still require an understanding of the nucleation properties of the system being studied in order to minimize the contribution of nucleation to the final result.

The following section will discuss homogeneous, heterogeneous, and secondary nucleation mechanisms and kinetics. This will be followed by a similar discussion of crystal growth. The reader is directed to the references cited above, and others, for detailed treatment of these phenomena.

4.2 NUCLEATION

Creation of new crystals (nucleation) can take place through a variety of mechanisms. Figure 4-2 summarizes the types of nucleation which can occur. Some of these are true nucleation (driven essentially only by free energy considerations); others are heavily dependent on imposed conditions (mixing and others).

The presence and width of a metastable zone, in which nucleation is not spontaneous, have been discussed in Chapter 2. The thermodynamic limit of the metastable zone is a locus of points known as the *spinodal curve*, where spinodal decomposition replaces nucleation and crystal growth as the phase separation. In typical industrial crystallization, nucleation (and release of supersaturation) occurs at much lower supersaturations than the spinodal curve.

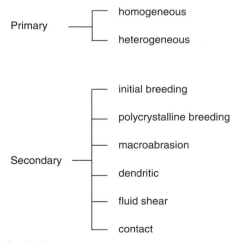

Figure 4-2   Mechanisms of nucleation.

4.2.1 Homogeneous Nucleation

As the cluster size increases, the specific surface (area/volume) decreases. For any given system, there is a critical stable cluster size at which the intermolecular force accompanying the addition of new molecules exceeds the force required to extend the surface area of the aggregate. If the aggregate is smaller than the critical size, the probability of its decomposition is great. Likewise, particles larger than the critical size are likely to survive and grow.

The classical nucleation theory of Volmer (1939), Nielsen (1964), and others assumes the addition of single molecules to a cluster until it reaches a critical size:

$$a + a = a_2$$
$$a_2 + a = a_3$$
$$\cdots$$
$$\cdots$$
$$a_{i-1} + a = a_i$$

During the cluster formation process, the free energy for nucleus formation is a balance between that for formation of the nucleus surface (positive) and that for phase transformation (negative). As the cluster becomes larger, the intermolecular forces of the particles within the cluster begin to predominate over the effect of the surrounding particles. Figure 4-3 was calculated by Johnson (2003) for β-carotene in 47 wt% tetrahydrofuran in water at $S = 2$, where $S = (C/C^*)$. The aggregation number (cluster volume) is used rather than the commonly used cluster radius because it relates directly to the cluster formation process taking place. At the critical cluster size Q^*, as shown in Fig. 4-3, the free energies for surface formation and phase transformation are in balance and the energy change for addition of another molecule becomes negative (the cluster is likely to survive and a nucleus is born).

In Fig. 4-3, Q^* is the critical aggregation number, ΔG_s is the surface excess free energy (between the surface and the bulk of the particle), ΔG_v is the excess free energy between a particle at $r = \infty$ and the solute in solution, γ_{12} is the surface tension, and ω is the molecular volume.

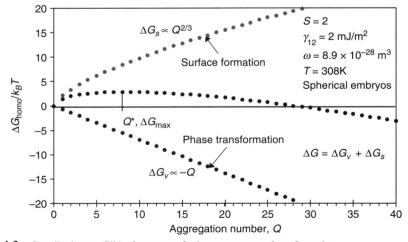

Figure 4-3   Contributions to Gibbs free energy for homogeneous embryo formation.

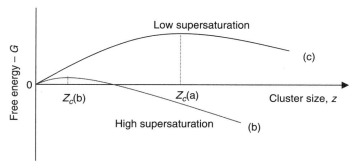

Figure 4-4   Effect of supersaturation on free energy of cluster formation.

As supersaturation is increased, the entropy of phase transformation is lowered, thereby lowering ΔG for formation of the condensed crystalline phase (nucleation), as shown in Fig. 4-4. In the β-carotene/THF/water example of Fig. 4-3, $S > 8$, the critical-sized embryo is about one or two molecules.

The rate of homogeneous nucleation of a sphere takes the following form, derived from an Arrhenius-type expression:

$$B_0 = A \exp\left[\frac{-16\pi\gamma_{12}^3\omega^2}{3k^3T^3(\ln S)^2}\right] \tag{4-1}$$

where B_0 = nucleation rate (number per m^3 per second), S = saturation ratio $[C/C^*]$, and v = molecular volume (m^3 per molecule). A is a prefactor containing solute diffusivity, supersaturation, molecular diameter and volume, equilibrium solute concentration, and Avogadro's number. The terms in the exponential reflect the energetics of phase change.

Derivations of Equation 4.1 are in the references cited above. One such location is Myerson (2002, pp. 45–46). Theoretical computation of A is given by Nielsen (1964). Predicting diffusion to the surface results in a value of $A \sim 10^{30}$. Measurements of A for lovastatin by Mahajan and Kirwan (1996) show Nielsen's value to be orders of magnitude too high, presumably because of the slow surface integration rate exhibited by many organic molecules.

4.2.1.1   *Limitations of Classical Nucleation Theory*

There are a number of oversimplifications in this theory, including the important one that classical thermodynamic considerations, which are based on averages of larger groupings of molecules, may not be applicable to small nucleating clusters. However, the most frequently discussed limitation is its basic presumption of the addition of one molecule at a time to the growing cluster. It is almost certain that the clusters or embryos aggregate and rearrange while building up to critical size. In practice, the entire process is also affected by local concentration fluctuations.

Especially with organics, which create complex crystal lattices held together by relatively weak forces, the rate of addition of molecules to the forming solid phase can exceed the ability of these molecules to orient themselves properly. This can easily result in a partially or totally amorphous structure in the solid. The frequency of this event can be minimized by providing seed particles with the desired final lattice structure. The authors suggest that seeded growth processes should be considered early in development.

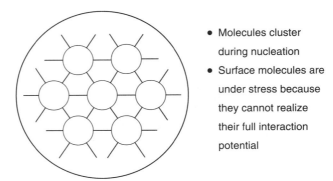

Figure 4-5   Stress on surface molecules in a cluster or crystal can alter its properties.

Other considerations in homogeneous nucleation relate to surface stresses in small crystals and to the effect of solute depletion in the immediate neighborhood of forming aggregates.

4.2.1.2   Ostwald Ripening

Clusters formed by the processes described above experience a phenomenon which is not a nucleation event but does affect the nature of the resultant crystal population. Figure 4-5 illustrates the fact that a cluster's surface molecules are under stress because they are insufficiently coordinated. This stress raises the pressure in the interior, increases the interfacial tension, and results in increased solubility of smaller entities, especially near the critical cluster size. This phenomenon is described by the Ostwald-Freundlich equation and leads to the process called Ostwald ripening, which is commercially important. Many industrial crystallization processes run age cycles, often with heating and cooling, to grow the larger crystals in the population at the expense of the smaller ones. Ostwald ripening is further discussed in Sections 4.3.1.7 and 4.4.

4.2.1.3   Localized Solution Concentration

Clusters which are approaching stability can be undermined by a localized reduced concentration, created by depletion from rapid nucleation. A schematic is shown in Fig, 4-6. Although this is a minor effect, in some systems it can increase the induction time for nucleation.

4.2.1.4   Nucleation of Polymorphs

Problems in the pharmaceutical industry related to the crystallization of polymorphs are discussed in Chapter 3. Both true polymorphs and solvates can form crystalline phases which are not the most thermodynamically stable entities in any given situation. Ostwald (1897) proposed a mechanism, now called the Ostwald Rule of Stages, in which the nucleating structure is not the final, most stable form, but rather that representing the smallest change in free energy. There is then a succession of small free energy reduction (nucleation) events taking place until the stable form is reached.

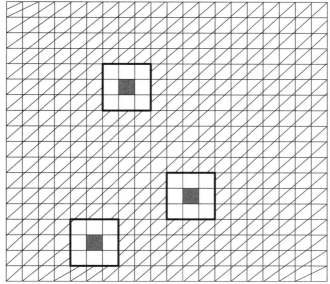

Figure 4-6   A schematic representation of the solution depletion in the vicinity of the aggregate being formed (the density of lines represents the concentration of solution).

Although the Rule of Stages is often operative in polymorphic systems, there are many systems in which it either does not apply or may apply only under certain supersaturation conditions. These criteria are described in Davey and Garside (2000, pp. 19–20).

4.2.2   Heterogeneous Nucleation

Heterogeneous nucleation can be primary or secondary, the former being caused by foreign particles and the latter by undissolved solute (usually crystallized from that solution).

Foreign substances in supersaturated solutions reduce the energy required for nucleation. Surface energy and particle geometry play a role. The form of Equation 4-1 is retained but the negative exponential is generally reduced, often significantly, by the changed surface tension and energetics. Figure 4-7 described the effect of wetting angle on the ability of the

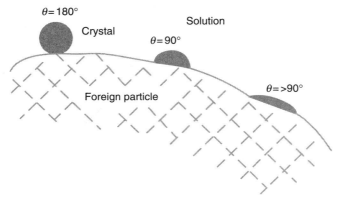

Figure 4-7   Nucleation on a foreign particle for different wetting angles.

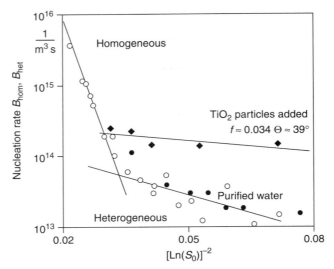

Figure 4-8 Measurement of nucleation kinetics for barium sulfate. Schubert and Mersmann (1996), by permission.

clusters to form on the foreign particles. A contact angle of $0°$ should result in spontaneous nucleation, but in practice the kinds of particles added, intentionally or otherwise, do not generally approach that condition.

Figure 4-8 shows data from Schubert and Mersmann (1996). Using a rapidly mixing device to study barium sulfate precipitation, they showed the transition from (secondary) heterogeneous to homogeneous nucleation rates (changed slope) as supersaturation was raised. They also showed that the heterogeneous nucleation rate was proportional to the particle surface area, not the number of particles. A significant additional result was their ability to further increase the heterogeneous nucleation rate by adding foreign particles of titanium dioxide.

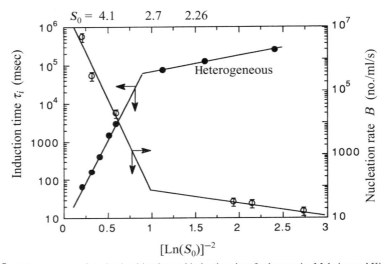

Figure 4-9 Measurement of nucleation kinetics and induction time for lovastatin. Mahajan and Kirwan (1996).

The same transition from heterogeneous to homogeneous nucleation occurs in the crystallization of organics. Figure 4-9 shows data from Mahajan and Kirwan (1996) for rapid crystallization of lovastatin, a pharmaceutical used for cholesterol reduction.

4.2.3 Secondary Nucleation

Secondary nucleation occurs when crystals of the solute are present. Less supersaturation is required to generate new particles under this condition. Referring to the types of secondary nucleation shown in Fig. 4-2, all of those are preformed crystals derived by various means from the parent, except for the case where molecular aggregates (clusters) are stripped from the surface. Except for pure breakage effects, as in milling, supersaturation is required in all types of secondary nucleation to satisfy the energetics.

Initial or dust breeding originates from crystallites formed on parent crystals during the growth period or broken off in storage. The crystallites exceed the critical size and are able to form new, fully formed crystals. As noted by Myerson (2002), if these are deemed undesirable for seeded batch crystallization, the crystallites must be removed by appropriate pretreatment (usually a solvent wash). The solvent power should be relatively low for the treatment step, as it is important to have enough volume to provide good washing of all the crystal surfaces.

Macroabrasion, usually called collision or attrition breeding, causes rounding of edges and corners. It can be an industrial problem, but it has a clear mechanism, and can often be controlled with suitable choices of mixing impellers and other parameters.

Needle breeding occurs at higher supersaturation from needle or dendritic growth which is released into the solution or slurry. Figure 4-10, from the classic paper of Clontz and McCabe (1971), shows how needle-like or spikewise growth can be produced at high

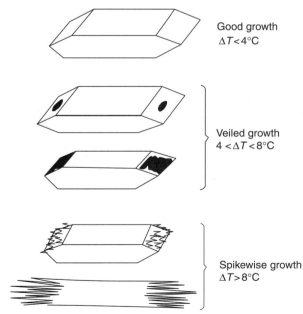

Figure 4-10   Effect of supersaturation on growth characteristics of $MgSO_4 \cdot 7H_2O$ (after Clontz and McCabe (1971)).

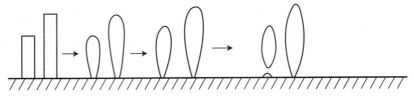

Figure 4-11   A schematic representation of dendrite coarsening.

supersaturation (usually unintentionally). Figure 4-11 shows the stages of coarsening which can occur with this morphology to produce nuclei.

Contact nucleation is the most common mechanism of secondary nucleation. Crystal-crystal-, crystal-impeller, and crystal-wall collisions are involved. Secondary nuclei arise from microabrasion (crystal surface damage) or ordered cluster removal by fluid shear forces, as noted above. Figure 4-12 shows that for a given substance, impeller speed and material of construction can both play a role.

Another type of nucleation, polycrystalline breeding, occurs at higher supersaturations from irregularly formed aggregates, but this is not a common problem.

4.2.3.1 Secondary Nucleation Correlation

For contact nucleation, the most common type of secondary nucleation, the combination of potential causes usually calls for a power law correlation of the type

$$B = kM^j N^k (\Delta C)^b \tag{4-2}$$

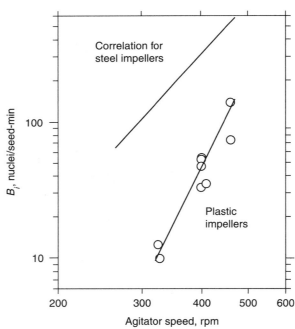

Figure 4-12   Effect of agitator speed on the secondary nucleation rate for steel and plastic impellers. Ness and White (1976).

where B is the nucleation rate in number per m^3 per second, ΔC is the concentration difference supersaturation $(C - C^*)$, N is the agitation rate (usually RPM or tip speed), and M is the suspension density.

The value of the exponent b is usually between 1 and 2.5, considerably lower than any predictions based on primary nucleation. The value of the exponent k is usually between 2 and 4. The value of j is generally close to 1, implying that collisions with the impeller(s) and other parts of the crystallizer are more important than crystal-crystal interactions.

4.3 CRYSTAL GROWTH

In any crystal, the orientation between any face and the crystallographic unit cell is defined by its Miller Indices, which are the inverse of intersections of the face with the crystallographic axes a, b, and c. For example, the surface (101) is parallel to the b-axis and cuts the a- and c-axes one unit cell from the origin. Good reviews of Miller Indices are provided by Mullin (2001, pp. 10–13) and Myerson (2002, pp. 35–36).

Crystal growth is a description of the linear velocity of a growing face, that is, the linear velocity perpendicular to that face. Growth velocity, in the classical view, is considered to be constant (size independent), according to the classical McCabe ΔL Law (McCabe 1929).

Crystal growth is envisioned as a growth of layers. As a molecule or cluster arrives at a surface, it must shed its layer(s) of solvent and bond with that layer. Figure 4-13 shows the bond configurations that are possible. Molecules A, B, and C are attached to, respectively, (A) the surface only (one attachment), (B) the surface and a growing step (two attachments), and (C) a "kink site" (three attachments). Energetically, (C) is most favorable for successful bonding to the surface, (B) is less favorable, and (A) is least favorable.

4.3.1 Crystal Growth Mechanisms

4.3.1.1 Continuous Growth

Some crystalline materials with high roughness contain enough kink and step sites on the surface to integrate essentially all approaching growth units with minimal complication. This is referred to as continuous growth. Under these conditions, the growth rate is linearly proportional to the supersaturation at all supersaturation levels.

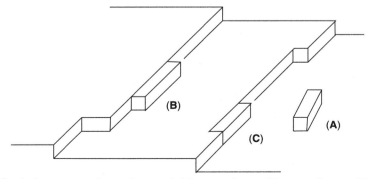

Figure 4-13 Surface structure of a growing crystal: (A) one attachment, (B) two attachments, (C) three attachments.

4.3.1.2   Growth by Two-Dimensional Nucleation

Many or most materials which are crystallized do not have enough roughness to exhibit continuous growth. Steps must be created to accept the incoming growth units. One mechanistic explanation for creation of these steps is two-dimensional nucleation at the surface. This involves the formation of nucleated circles on the flat crystalline surface, similar to the three-dimensional nucleation described earlier in this chapter. As illustrated in Fig. 4-14, molecules are being adsorbed onto, desorbed from, and diffusing on the crystal surface.

There are several possible mechanisms for completion of the growth layer after surface nucleation. The mononuclear model envisions infinite spreading velocity of the surface nuclei. In this model, the growth rate of each face is proportional to its area, diametrically opposite from the observed behavior, in which the largest faces are the slowest-growing. The polynuclear model assumes zero spreading velocity. This predicts a global maximum growth rate at intermediate supersaturation. The data of Mahajan and Kirwan (1996) fit this model. The birth and spread model allows a finite spreading rate and nucleation at any location on the surface. It is sometimes used for correlation of data.

4.3.1.3   Screw Dislocation (Spiral Growth)

All of the above models assuming two-dimensional surface nucleation predict crystal growth rates at low supersaturation which are much lower than those observed in practice. A possible solution to this problem was put forward by Frank (1949), who proposed a self-generating step creation process involving a screw dislocation. The model was further formalized by Burton et al. (1951) and became known as the Burton-Cabrera-Frank (BCF) model of crystal growth. Detailed discussion of the BCF model can be found in Ohara and Reid (1973) and Nyvlt et al. (1985).

Self-generating step creation, such as that provided by the BCF model, involves inherently more rapid growth kinetics because of the more favorable energetics for addition of a molecule or cluster to the structure.

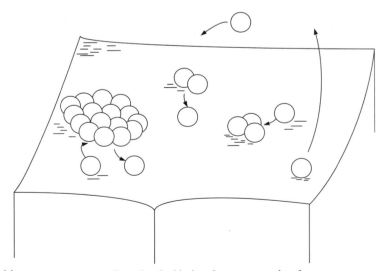

Figure 4-14   Formation of a two-dimensional critical nucleus on a crystal surface.

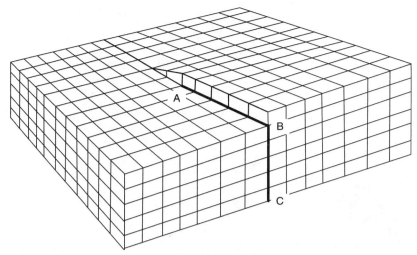

Figure 4-15 A screw dislocation in a simple cubic crystal. AB, BC are dislocations. The screw dislocation AD is parallel to BC (D is not visible).

Figures 4-15 and 4-16 illustrate the formation of a screw dislocation. Excellent microphotographs of spiral crystal growth are shown in Fig. 4-17 (Mullin 2001, pp. 221–222). In Fig. 4-17, the lower left photo is of an organic substance (C_{36}, normal alkane).

Dislocations can occur within crystals as a result of stresses in crystal growth on seeds and around surface nuclei and inclusions. Spiral and other self-perpetuating growth can promote growth rate dispersion (different growth rates on different crystals of the same material) because each crystal is responding to the structure of its own unique dislocation(s).

4.3.1.4 Effect of Impurities and Solvents

Impurities can affect the rates of nucleation and crystal growth, often (or usually) with an effect disproportionate to the amount involved. Part per million levels of impurity have been shown to profoundly influence the size, shape, and rate of growth of some industrially important compounds. Entities as common as table salt are often modified intentionally by producers seeking desirable properties.

One "impurity" always present in solution crystallization is the solvent. It is present, of course, in larger quantity, and in addition to the problem of (often multiple) solvated entities, the solvent molecules must rapidly diffuse away from the growing crystal faces.

The interfacial layer of molecules on a growing crystal surface is considered by most to be intermediate in structure between fully crystalline and amorphous layers. Impurity

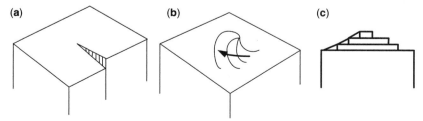

Figure 4-16 Development of a growth spiral starting from a screw dislocation: (a) start, (b) later, (c) final spiral.

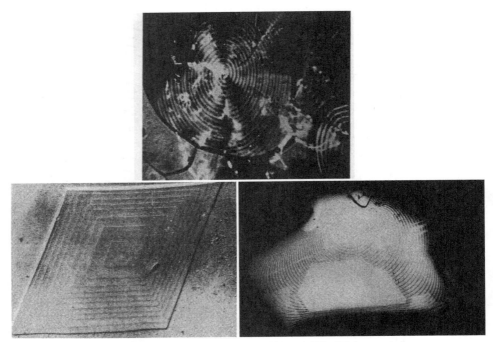

Figure 4-17   Examples of spiral crystal growth. Mullin (2001), pp. 221–222, by permission.

molecules and clusters from this layer are in competition with the desired compound for adsorption sites on the surface. Figure 4-18 illustrates adsorption of impurities in the kink sites, steps, and ledges shown in Fig. 4-13.

Equilibrium incorporation of impurities can take place if they occupy spaces meant for other solute molecules (substitutional defects), actual open spaces (vacancy defects), or smaller interstitial spaces. Nonequilibrium impurity uptake (in inclusions, dislocation planes, permanently adsorbed entities, and impurities left in adhering mother liquor) is common at the higher rates pushed in industrial processes, because the solute molecules attracted to the surface at a high rate can physically slow down reverse diffusion of the impurity. Additionally, any adsorptive attraction of the impurity to the crystal surface further exacerbates this effect.

Tailor-made additives are sometimes used to inhibit growth or activity on certain faces. These are structurally designed molecules which fit these surfaces in a lock-and-key arrangement (Davey et al. 1991; Weissbuch et al. 1995).

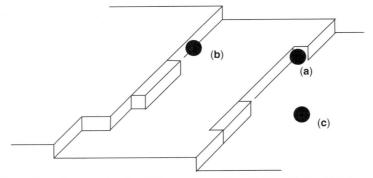

Figure 4-18   Distinct adsorption sites for additives and impurities: (a) kink, (b) step, (c) ledge.

Table 4-1 First-Order Growth Rate Constants (Fluidized Bed versus CSTR Crystallizer System)

	$k \times 1000$: gm/hr, cm^2, gm/cm^3		
	Fluidized bed	CSTR	
Crystallizer temp. (°C)	No nucleation suppressant		Nucleation suppressant added
25	40		2.2
35	70	16	2.3
41	90		2.3
46			4.1

Impurities, added or unintentional, can have a major effect on rates of nucleation and crystal growth. Table 4-1 shows the effect of an impurity, structurally similar to the crystallizing solute, added to an all-growth crystallization (separation of stereoisomers, Examples 7-6 and 11-6). The data for a continuous stirred tank (CSTR) operation show a sevenfold decrease in the first order growth rate constant as a result of addition of this impurity to prevent nucleation of the undesired isomer.

It is clear, in this case, that the suppressant impurity "poisons" the growth surface, presumably by adsorption (CSTR with suppressant versus no suppressant). A higher growth rate in the fluidized bed versus the CSTR implies considerable diffusional resistance (higher fluid slip velocities in the fluidized bed result in improved growth rate).

Impurities are known to profoundly affect crystal morphology. The texts referred to near the beginning of this chapter offer some structural models of impurities that interface with the growth of specific faces. An example from the authors' experience is shown in Fig. 4-19. The three microphotographs are of the same crystal form of a compound "spiked" with small amounts, respectively, of each of three known impurities normally present in much smaller amounts in the mother liquors from which growth is taking place.

4.3.1.5 Mass Transfer Limited Growth

Crystallization from solution often has to overcome serious limitations in bulk diffusion, particularly in large-scale industrial crystallizers. These limitations have been addressed in a number of formats.

The best-known of the early developments addressing mass transfer effects on the BCF model was that of Chernov (1961), setting diffusion of solute through a boundary layer as the rate-limiting step. Other, more complex descriptions taking both surface and bulk diffusion into effect have been presented by Bennema (1969) and Gilmer et al. (1971).

An effectiveness factor has been presented by Garside (1971) as the ratio of the actual growth rate to the one that would occur if the interface were exposed to the bulk conditions. This is also presented as a Damkoehler number, but it is different from the Damkoehler numbers described in Chapter 6, which relate to mixing time/process time ratios.

Diffusion-limited crystal growth processes require attention to mixing and other aspects of fluid flow around the crystals. These issues will be discussed in more detail in Chapter 6. As an example of the effect of hydrodynamics on crystal growth kinetics, refer to Table 4-1 in the above discussion on effect of impurities. Comparing the growth rate in a fluidized bed

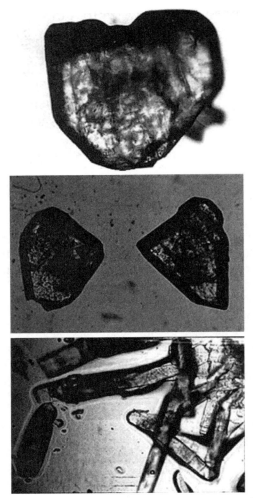

Figure 4-19   Low levels of impurity can affect crystal morphology: three views of the same compound grown under the same conditions with different known impurities.

with that in a CSTR, both with no added impurity, results in more than a fourfold improvement in the fluidized bed under the same process conditions. The most credible explanation for this improvement is the increased slip velocity of the fluid around the fluidized particles. The positive effect of higher temperature on growth rate, in this case, is probably caused by a reduction in resistance to both mass transfer and surface integration.

4.3.1.6   Kinetic Order for Crystal Growth

The introductory section of this chapter cites texts showing full development of the crystal growth models described above. Those models describing continuous growth, two-dimensional nucleation, and screw dislocation (BCF and variants), with or without diffusional limitation, predict values of kinetic order (exponent r) between 1 and 2 in the expressions

$$G = k_G(C - C^*)^r \tag{4-3}$$

and

$$R_G = k_{GM}(C - C^*)^r \tag{4-4}$$

where G is linear crystal growth velocity and R_G is the mass growth rate per unit surface area. The growth rate constants k_G and k_{GM} are temperature dependent.

The experience of the authors is in general agreement with a growth kinetic order between 1 and 2. Indeed, many pharmaceutical compounds exhibit simple first-order growth with no complication. There are some exceptions, however, and some of these come without explanation. Figure 4-20 shows (relative) growth rate constants (k_{GM}) for a cooling, all-growth crystallization of a compound solubilized by the presence of an acid (acid salt more soluble than the free base). As can be seen, the kinetic order is 2.5. The compound in Fig. 4-20 is one of several that have shown some level of higher-order kinetics.

4.3.1.7 Size-Dependent Growth and Growth Rate Dispersion

Models of crystal growth generally assume a size-independent growth rate (obedience of McCabe's ΔL Law, as noted in Section 4.3). Many exceptions have been found to this rule, with the rationale for them matching the situation.

1. As discussed in Section 4.2.1.2, very small particles (near the 1 micron range) are more soluble than larger ones because of unsatisfied outer bonds and resultant internal pressure. Thus, these very small particles grow more slowly because of a

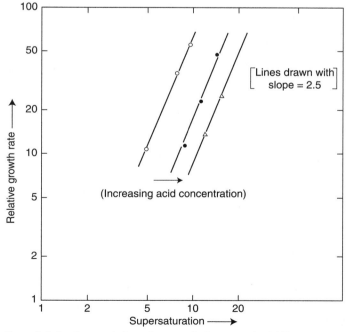

Figure 4-20   Shown is 2.5-order growth for a pharmaceutical compound solubilized by acid.

reduced supersaturation driving force. As will be further discussed in Section 4.5, with aging or temperature cycling, these particles can disappear from Ostwald ripening.

2. Even in sizes considerably larger than the micron/submicron range, however, smaller crystals have been shown to grow at a lower rate than larger ones. The mechanisms include expected size dependence based on a higher expectation of surface dislocations (larger particles have higher mechanical stresses, collision energy, and impurity inclusions) and bulk diffusion effects (higher slip velocity for larger crystals). Rate expressions incorporating these effects have been summarized by Randolph and Larson (1988).

Situations 1 and 2 are referred to as size-dependent growth. Another reason for variation of growth rates in a crystal population is growth rate dispersion. This dispersion occurs because of differences in individual crystals. These differences are based on (a) inherent properties of each nucleus, (b) fluctuations with time of each nucleus, or (c) expected differences between crystals growing from (differently sized) dislocations by the BCF mechanism described in Section 4.3.1.3 or by others.

Size-dependent growth and growth rate dispersion do affect the ultimate size distribution in industrial crystallizers. To a large extent, measurement of growth rates with large numbers of crystals in suspension adequately compensates for individual variations within the crystal population.

4.3.2 Measurement of Crystal Growth Rate

Measurements of both nucleation and growth rates are of great assistance in the development of a crystallization process. Since nucleation rates are strongly dependent on many factors (impurities, hydrodynamics) which might be changing during evolution of the process, the early measurements are usually those of the growth rate. Additionally, since the authors of this book encourage the use of growth-based processes if possible, these measured growth rates can then be used to assist at all levels of scale-up.

Newer, in-line techniques are making it possible for easy measurement of the growth rate (and the nucleation rate under these conditions) in small-scale experiments. However, the techniques used in more traditional measuring systems still provide perhaps the most usable data for scale-up.

Myerson (2002), Mullin (2001), and Mersmann (2001) provide excellent descriptions of methods for crystal growth rate measurements. These methods involve measurements of either single crystals or suspensions. Much information can be gained from the traditional technique of measuring ("grab" samples or in-line) solute concentration versus time in batch crystallization on a seed bed. Initial and later slopes on such a plot can provide multiple data points of growth rate versus supersaturation.

4.3.2.1 A Useful "Workhorse" Measurement System

The authors have found that semicontinuous crystallization in a small fluidized bed (typically 1-inch diameter), in a system shown schematically in Fig. 4-21, provides excellent, high-quality data usable through scale-up to full-sized manufacturing processes. The "dissolver" vessel is maintained as a saturated slurry by using an excess of solid and an internal filter (examples: sintered vessel bottom or sintered "candle" filters).

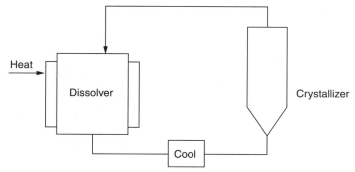

Figure 4-21   Fluidized bed crystallizer growth rate test apparatus.

Optional features of this system are as follows:

- The fluidized bed crystallizer typically has a screen support (as in a commercial chromatography column) but can be operated without one by use of a tubing clamp on the column inlet line. (The clamp must be closed quickly whenever the pump is stopped.)

- If the crystallizer is run at ambient temperature, it is not necessary to use a jacket on the column. However, even at ambient internal temperature, the crystallizer shows reduced (or totally eliminated) internal wall growth if a jacket is used (even without cooling fluid).

- The crystallizer exit stream can have an optional bypass through a line filter (typically sintered glass), which is used for the first 10 minutes or so. This filter can collect fines present in the feed solid caused by the breakup of agglomerated, inherently smaller particles. The weight of solids collected in the line filter is deducted from the initial seed weight when calculating the growth rate. The line filter, in most cases, is not needed, but if one is available, it is useful to check it out when embarking on a set of experiments using new seed material.

- Typically, using about 5 gm of monosized seed (usually a sieve fraction), and harvesting after several hours of operation, can produce an accurate and reproducible growth rate measurement. The effects of supersaturation (including the data in Fig. 4-20), impurity level in the mother liquors, and absolute temperature are easily determined.

The final dry weight of the product can be used to measure the growth rate, which can be calculated assuming constant surface area/mass:

$$\text{Growth Rate (\%/hr)} = [100\ln(W_f/W_o)]/\theta \tag{4-5}$$

where W_f/W_o = final seed mass/initial seed mass and θ = time (hr).

This measured growth rate incorporates the surface area and supersaturation terms in the kinetic expression:

$$\text{Growth Rate (\%/hr)} = 100\, k_{GM}A(C - C^*)^r \tag{4-6}$$

where k_{GM} is the constant from Equation 4.4 and A is the total surface area.

If sufficient growth has taken place that change in surface area/mass should be taken into account, then surface area/mass (for three-dimensional growth) is proportional to

$(W_f/W_o)^{2/3}$ and

$$\text{Growth Rate } (\%/\text{hr}) = (300/\theta)[1 - (W_o/W_f)^{1/3}] \qquad (4\text{-}7)$$

The *initial* growth rate, which represents the productivity of the initial seed before reduction in surface area/mass, may be particularly useful if that seed is to be used in a scaled-up operation:

$$\text{Initial Growth Rate } (\%/\text{hr}) = (300/\theta)[(W_f/W_o)^{1/3} - 1] \qquad (4\text{-}8)$$

4.3.3 Crystal Population Balance

The population balance approach to measurement of nucleation and growth rates was presented by Randolph and Larson (1971, 1988). This methodology creates a transform called population density $[n(L)]$, where L is the characteristic size of each particle, by differentiating the cumulative size distribution N versus L, shown in Fig. 4-22, where N is the cumulative number of crystals smaller than L. Per unit volume, the total number of particles, total surface area, and total volume/mass are calculated as the first, second, and third moments of this distribution.

Randolph and Larson showed that for a continuous crystallizer with a (perfectly) mixed suspension, (perfectly) mixed product removal (MSMPR) crystallizer,

$$n = n^o \exp(-L/G\tau) \qquad (4\text{-}9)$$

where n^o is the population density at $L = 0$ (i.e., the nuclei), G is the linear crystal growth rate, and τ is the residence time in the crystallizer.

Thus, running an MSMPR experiment and plotting ($\ln n$ versus L) should result in a straight line with slope $(1/G\tau)$, as shown in Fig. 4-23. Since τ is known, the linear growth rate is known under the crystallization conditions. The intercept n^o is converted to

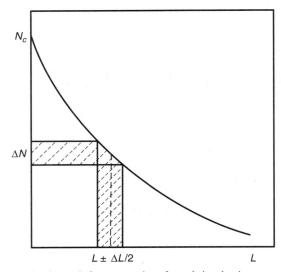

Figure 4-22   Size and number intervals for computation of population density.

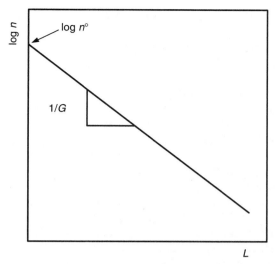

Figure 4-23   Graphic representation of the population balance of the MSMPR crystallizer.

the nucleation rate B^o by the relation

$$B^o = n^o G \tag{4-10}$$

The line from MSMPR experiments is usually straight but often has variants. These are shown in Fig. 4-24.

While this methodology has been shown to be very useful for many crystallization processes, it is subject to size dispersion and the other nonlinear effects of Fig. 4-24, and has the additional need to assure (perfect) mixing up to and including the product exit line. Use of large amounts of material is often impossible for pharmaceutical products in the early stages of development, so this technique has only limited use in our industry.

Type	Appearance	Cause
Linear	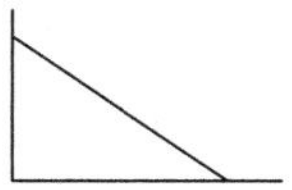	MSMPR formation is obeyed
Concave upward	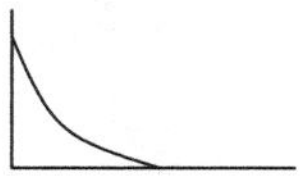	(1) Size-dependent growth (2) Growth rate dispersion (3) Agglomeration (4) Classification
Local maxima	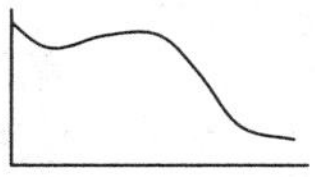	(1) Agglomeration with breakup (2) Classification

Figure 4-24   Summary of semilogarithmic population density plots and potential causes.

4.4 NUCLEATE/SEED AGING AND OSTWALD RIPENING

Ostwald ripening, the disappearance of very small crystals because of an increase in solubility, has been discussed in Sections 4.2.1.2 and 4.3.1.7 Despite the fact that the concept applies most directly to particles ≤ 1 micron in size, it has considerable commercial significance because of its participation in the aging of seed to increase the size, narrow the size distribution, and improve the crystallinity of particles in a crystal growth operation. Mullin (2001, pp. 320–322) contains an excellent discussion of this phenomenon.

Seed particles are produced in many ways, and their production and use are discussed in many places in this book. Many seed generation processes, including contact nucleation (discussed earlier in this chapter) and milling, produce particles with more physical stresses and wider size distributions than desired for good growth operation. Seed aging is frequently utilized, either after seed addition or after initial nucleation, to minimize the smallest and most stressed particles in the population. In general, these particles are characterized by higher solubility, the smaller ones by a combination of the surface stresses and Ostwald ripening, and some larger ones by stresses alone.

The aging, or ripening, of seed (and sometimes of the final product) can take place at a single temperature through normal dynamic equilibrium processes, but it can be significantly

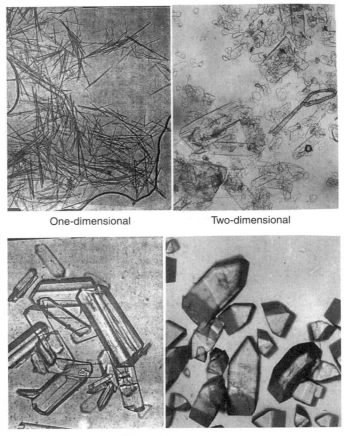

One-dimensional   Two-dimensional

Three-dimensional

Figure 4-25   Typical crystal growth shapes.

accelerated by higher temperatures and, in particular, by temperature cycling. The overall process of seed aging carried out in this manner is thermal digestion. Several such processes are part of some of the examples in this book.

4.5 DELIVERED PRODUCT: SIZE DISTRIBUTION AND MORPHOLOGY

Depending on the relative growth rates of the faces of a given crystalline product, it can be considered to grow in a one-, two-, or three-dimensional form. Pharmaceutical products from these respective growth types are shown in Fig. 4-25.

Crystallization processes using either added or nucleated seed can be designed to manufacture product with a desired product size distribution. The following steps must be taken to achieve this goal:

- Supersaturation and mixing environments must be designed to minimize or eliminate unexpected nucleation beyond that required for seed. A seed age step must be added if needed.

- Seed or nucleate size distribution must be controlled and reproducible.

The ultimate particle size distribution can be calculated by simple geometry. A simple calculation is shown in Table 4-2 for growth on 5 micron cubes at different seed/nucleate levels, for three-, two-, and one-dimensional growth. As noted in the note at the bottom

Table 4-2   Effect of Extent of Seed or Initial Nucleation on Final Crystal Particle Size

Seed or nuclei $\longrightarrow$	Product particle dimensions		
	Length (5 μm)	Width (5 μm)	Thickness (5 μm)
Product: A. Cubic particles (three-dimensional growth)			
Seed level			
0.5%	29 μm	29 μm	29 μm
1%	23 μm	23 μm	23 μm
5%	14 μm	14 μm	14 μm
10%	11 μm	11 μm	11 μm
Product: B. Flat plates (two-dimensional growth)*			
0.5%	71 μm	71 μm	5 μm
1%	50 μm	50 μm	5 μm
5%	22 μm	22 μm	5 μm
10%	16 μm	16 μm	5 μm
Product: C. Needles (one-dimensional growth)*			
0.5%	1000 μm	5 μm	5 μm
1%	500 μm	5 μm	5 μm
5%	100 μm	5 μm	5 μm
10%	50 μm	5 μm	5 μm

*Very thin particles are unlikely to survive attrition in the tank.

of the table, some resultant particles cannot be expected to survive under normal conditions of mixing or fluid dynamic conditions, because in real industrial crystallization processes, nucleation and crystal growth inevitably occur simultaneously. To better understand the impact of nucleation and crystal growth and control over the final particle size distribution, integrating the crystallization kinetics modeling with PAT technology has been proven helpful (Togkalidou et al. 2004; Woo et al. 2006; Zhou et al. 2006). Regardless of the techniques, the keys are seeding and maintaining superssaturation to maximum crystal growth.

Chapter 5

Critical Issues in Crystallization Practice

5.1 INTRODUCTION

A crystallization process for a specific compound will be governed by the compound's inherent properties for nucleation and growth. Both morphology and polymorph formation are also species specific. The resulting crystal physical and chemical attributes will also depend on properties of the surrounding environment such as solvents and temperature and the rate of generation of supersaturation. For organic compounds, these critical properties vary over exceedingly wide ranges that are caused by their structural differences. The reader is referred to Chapters 2, 3, and 4 for a discussion of these properties.

Direct scale-up from an undeveloped laboratory procedure may result in a different product from the expected one. These differences may include smaller mean particle size, wide and/or bimodal particle size distribution (PSD), needles or plates with difficult downstream processing properties, and lack of consistent results.

The purpose of this book in general, and of this chapter in particular, is to highlight the complex interactions between the inherent properties noted above as well as the possibilities for their manipulation to achieve the desired outcome. The objective is to present developmental, design, and scale-up guidelines that may assist in evaluating and manipulating the inherent properties of the compound and its crystallization environment with the goal of achieving a more robust and scalable process than might be achieved by direct scale-up of a laboratory procedure.

Most examples in this book illustrate successful manipulation to achieve particular goals. In some cases, however, the natural growth and/or morphology has not been modified sufficiently by the corrective action to achieve success. Example 9-3 is a case in which satisfactory growth could not be achieved. For difficult situations such as these, downstream problems can be minimized by optimization of process variables (Johnson et al. 1997; Kim et al. 2005).

5.2 NUCLEATION

In the absence of control of supersaturation, nucleation will usually predominate. Nyvlt et al. (1985, p. 36) notes that "the primary requirement in control of the crystallization process is thus control of the number of crystals formed". As discussed below, this control may be exceedingly difficult to achieve on any scale and may be particularly difficult to reproduce on scale-up, often because of mixing issues.

A specific crystallization can be dominated by either nucleation or growth. Which of these does dominate depends on the methods of control of critical variables, the amount of seed, the size distribution and surface qualities of the seed, and the environment in which supersaturation is created, as well as the specific nucleation and growth rate properties of the compound.

Both nucleation and growth almost always proceed simultaneously. In general, nucleation will dominate when supersaturation, either local or global, is close to or greater than the upper limit of the metastable region. Growth can dominate at low supersaturation and in the presence of a sufficient crystal surface area.

Three different mechanisms for nucleation may occur in any crystallization operation, as discussed by Mersmann (2001, pp. 45ff.): homogeneous primary, heterogeneous primary, and activated secondary. Industrial crystallizers are usually operated under conditions of the last two simultaneously. For simplicity, nucleation as used in this discussion refers to any or all three of the mechanisms that may be important in a particular crystallization.

A nucleation-dominated process may be chosen in order to produce fine particles for a product attribute such as bioavailability for a pharmaceutical with low water solubility. The reader is referred to Examples 9-5 and 9-6 for discussion of a process for creating fine particles. Nucleation-dominated processes may be difficult to operate in stirred vessels for the reasons outlined in Section 5.2.1 below and may require intense in-line mixing, as presented in the examples.

5.2.1 Nucleation Issues

In some cases, a process that is dominated by nucleation can result in an acceptable process outcome. The process may appear satisfactory in a laboratory-scale operation. However, there are several potential problems with this type of process for laboratory and pilot plant operation or when scaling up for manufacturing, including:

- fine crystals and/or wide PSD
- high surface area
- low bulk density
- the risk of large batch-to-batch variation
- occlusion of solvent and impurities
- agglomeration/aggregation
- lack of control of hydrates, solvates, and polymorphs

In addition, scale-up of a nucleation-dominated process is difficult to predict, unless the generation of supersaturation is well controlled. The difficulties associated with stirred-batch crystallization scale-up relying on nucleation were highlighted by Nyvlt (1971, p. 111):

> *Its disadvantages are the comparatively large amount of manual labor involved, and as a rule, a not very high degree of reproducibility of product quality. The product quality suffers mainly because the rate of supersaturating in the stage following introduction of a new batch is excessive, so that a large number of nuclei form in a rather uncontrolled manner. These cannot subsequently grow into large crystals. The product is, therefore, fine-grained, with the attendant difficulties of filtering, centrifuging, and washing. It retains an appreciable amount of impure mother liquor, dries badly, and tends to cake in storage.*

A principal reason for the difficulty in controlling local surpersaturation and nucleation is the differences in the local energy dissipation rates within a stirred vessel.

With an extremely high nucleation rate, for example, one with a characteristic time in milliseconds, the degree of control required to prevent formation of the large number of nuclei, as noted by Nyvlt above, may not be achievable in a stirred vessel in part because of the inherent properties of the compound and/or because of mixing scale-up issues (the most common being the geographical distribution and magnitude variation of the energy dissipation rate). Extreme examples of this are found in ionic reactions to precipitate inorganic salts. Organic acid-base reactions can also generate nuclei at high rates. An in-line type of crystallization approach may be required for antisolvent and/or reactive crystallization, both of which may be dominated by nucleation in the absence of strategies to promote growth. The processes referred to above for creating fine particles (Examples 9-5 and 9-6) are carried out using in-line crystallizers, in these cases variants of impinging jet mixing devices.

Nucleation must also be minimized by tight control of supersaturation in processes involving resolution of optical isomers. In some cases, nucleation must be avoided to prevent the formation of undesired polymorphs.

5.2.2 Nucleation Rate

As discussed in Chapter 4, the nucleation rate is both species specific and a function of the supersaturation ratio. The relation between nucleation rate, growth rate, and particle size as a function of the supersaturation ratio is illustrated qualitatively in Fig. 5-1. The actual rate and supersaturation characteristics, such as metastable zone width, are system specific and can vary over wide ranges. In practice, it has been observed that the nucleation rate may vary from milliseconds to hours, and the metastable zone width may vary from less than 1 mg/ml to tens of mg/ml.

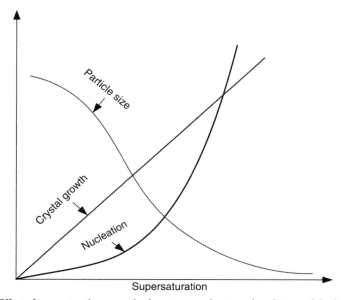

Figure 5-1   Effect of supersaturation on nucleation rate, growth rate, and nucleate particle size.

In addition, for a specific compound, the nucleation rate is also dependent on the solvent(s) system, impurities, and mixing. These factors combine to cause the difficulties that are often encountered in controlling a nucleation-based crystallization process, especially on scale-up.

The variation in supersaturation range, as indicated by the variability in width of the metastable region, can play a critical role in the operability of a process. An example of this type of variability is described in Example 11-1. In this process for the crystallization of an antibiotic, a high degree of supersaturation can be achieved and maintained for several minutes ($S \sim 15$) in an all-aqueous system allowing completion of sterile filtration at this high degree of supersaturation (high concentration). The subsequent addition of 3% acetone, however, effectively increases the nucleation rate and narrows the width of the metastable region (as defined in Section 2.2). Crystallization is then completed by the addition of sufficient acetone to achieve the desired yield.

As noted above, nucleation rate differences can be extreme within the same crystallization process. With small changes, a nucleation-based process can be made to run predictably when the nucleation rate is low (characteristic nucleation time is tens of minutes or hours), and can be essentially an all-growth process. At the other extreme, nucleation control can be virtually impossible when high rates are encountered (characteristic time of seconds or less).

5.3 GROWTH

When crystal growth is desired, tight control of several variables is almost always required. The outcome is species dependent, however, and in some cases growth may not be achievable because of the inherent growth rate or morphology. Nucleation followed by growth on a large number of nuclei may then predominate, thereby limiting the ultimate particle size, as discussed below in Section 5.5. As noted in many examples in this book, the type of crystallization equipment selected can play a key role in determining the success or failure of the operation.

Processes that require and/or benefit from operation under conditions in which growth predominates include

- required product purity
- large three-dimensional crystals for easy downstream operations: filtration, washing, drying
- predictable dry solid flow characteristics
- control of polymorphs
- resolution of optical isomers

5.3.1 Growth Issues

A growth-dominated process may have several advantages, including

- larger mean particle size
- lower surface area
- higher bulk density
- improved solvent/impurity rejection
- improved control of hydration, solvation, and polymorphs

- decreased sensitivity to mixing
- improved reproducibility on scale-up
- adaptability to continuous operation

All scales of operation are dominated by the effects of supersaturation, although the outcome is usually more critical on scale-up to pilot plant and plant operations. The local and global supersaturation ratios that are experienced over the course of a crystallization operation are critical because they determine the balance between nucleation and growth, not only at the onset of crystallization, but throughout the course of a batch or semibatch operation. This balance, in turn, determines the resulting physical properties and, in many cases, the distribution of chemical impurities between the crystals and the liquors.

5.3.2 Growth Rate and Growth Characteristics

Inherent crystal growth rates are system specific and can vary over several orders of magnitude for different compounds. Measurement techniques for growth rate are presented by Myerson (2002, pp. 58ff.) and Mersmann (2001, pp. 220ff.). A laboratory procedure for measuring the growth rate in a fluidized bed—chosen to minimize nucleation—is presented in Chapter 4. Readers can also read Example 7-2, which discusses crystal growth impact on particle size.

The factors influencing the crystal growth rate of a specific compound are discussed by Myerson (2001, pp. 93ff.). In addition to molecular structure and the solvent system, the growth rate can be greatly modified by the presence of dissolved impurities that may either compete for growth sites or block these sites. As with the nucleation rate, these differences can be so extreme as to make growth impractically slow, or at the opposite extreme, the rate may be sufficient to achieve an essentially all-growth process with careful control of supersaturation and growth area. In practice, it has been observed that the release rate of supersaturation by crystal growth can vary from less than a second to several days.

Supersaturation has an important effect on the growth rate, as shown in Fig. 5-1. As can be seen, the increased growth rate that can be achieved at higher supersaturation may come at the expense of increased nucleation, leading to broader PSD and possibly a bimodal distribution.

The effect of supersaturation also can depend on the solvent system, as shown in Fig. 5-2 for hexamethylene tetramine (Davey et al. 1982).

In the design of a crystallization process, therefore, the balance that is achieved between nucleation and growth rates is critical to particle size under the operational constraints of equipment and facilities. The supersaturation ratio can be controlled to limit nucleation in order for growth to predominate. This becomes increasingly difficult at lower inherent growth rates since it will extend the batch time cycle substantially and because the nucleation rate becomes more critical at lower growth rates.

Solvent and impurity effects must also be considered. Solvent effects are important and may play a key role in inclusions and in affecting the width of the metastable zone, as discussed in Example 11-1. However, variations in impurity composition can have more influence and can dominate the course of crystallization in many ways.

Impurities can

- slow the nucleation rate, leading to high supersaturation before oiling out (see Section 5.4 below) and/or crashing out
- retard or stop growth
- co-crystallize and/or form solid solutions (see Chapter 2)

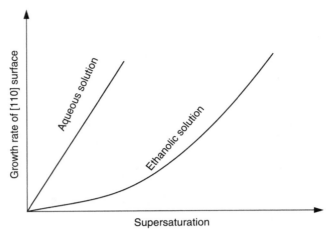

Figure 5-2   Comparison of the growth rate of hexamethylene tetramine crystals as a function of super-saturation in aqueous solution and in ethanol solution. (Reproduced with permission from Davey et al. 1982.)

Experimentation is required to evaluate these effects. The most useful experiments utilize spiking with known impurities when they can be isolated for this use. However, as is often the case, the number and possibly the low concentration of impurities often make this impractical. An experimentally simpler method is to recrystallize the compound with and without spiking of the mother liquors obtained from the process isolation. Differences in nucleation and growth may readily be observed by comparing photomicrographs of the resulting crystals. Both size and shape can be expected to be affected. If no significant differences are observed, the impurities from the process may not cause any nucleation or growth changes and the inherent properties of the compound may be assumed to prevail.

Ideal steps in determining growth potential include the following:

- purification to the highest possible extent (using chromatography if necessary)
- selection of a solvent with solubility <50 gm/liter and some dependence on temperature
- preparation of a clear solution with low supersaturation
- aging of this solution with minimal or no mixing in the presence of some seeds, and/or
- subjecting the solution to heat/cool cycles (some fines dissolve during each heating cycle, and some growth may occur on slow cooling)

This procedure may show that growth is possible. A growth rate can then be determined by various methods, including the fluid bed method described in Chapter 4 using the crystals from the heat/cool experiments as seed.

5.4   OILING OUT, AGGLOMERATION/AGGREGATION

The most important property of a compound from a crystallization point of view may be its inherent growth characteristics. The critical question is: *will it grow*? A method to evaluate growth potential is outlined in the previous section. Additionally, in an actual crystallization operation, in order to be crystalline, nucleation must occur and the resulting nuclei are then assumed to grow to some limit in the nucleation phase. The question here is whether, after reaching this nucleation limit, further conditions of growth will or will not result in additional growth. While the authors have found that most compounds will continue to grow to some limit, depending on many process and inherent factors, there are some that do not exhibit significant growth beyond the 5–10 micron range.

A substantial-sized subgroup of those compounds which do not exhibit typical crystal growth, in which a repeated lattice grouping or crystal structure, is so difficult to achieve that the compound resembles a liquid as it emerges from solution. This is the phenomenon of oiling out, which is often accompanied by the additional complication of agglomeration/ aggregation.

It is perhaps helpful to begin this discussion with the consideration of oiling out since it may be the first event in the pathway of crystallization or, in extreme cases, the operation may end with an oil, gel, or intractable gum or tar (Bonnett et al. 2002).

5.4.1 Oiling Out

Oiling out can be a critical factor in crystallization by any of the methods of creating supersaturation and becomes increasingly possible under several conditions, including

- high supersaturation
- rapid generation of supersaturation
- high levels of impurities
- presence of crystallization inhibitors even at low levels
- absence of seed
- inadequate mixing (high local supersaturation)

Oiling out is species dependent and may be more prevalent for low-melting compounds, although in many respects it resembles the solidification of high molecular weight compounds such as polymers. As discussed in Chapter 2, oiling out can be considered as a spontaneous phase split into two liquid phases. On a Gibbs free energy–composition diagram, it represents a system in which the overall composition has exceeded the critical composition for spinodal decomposition.

A mechanism for oiling out can be postulated as follows: When supersaturation is achieved rapidly such that the concentration is beyond the upper metastable limit—as can often be the case in a nucleation-based process—the substrate is forced to separate into a second phase by the creation of the resulting high solution concentration. However, crystallization is delayed by a slow crystallization rate. This combination may result in the creation of a nonstructured oil or possibly an amorphous solid. The rates of phase separation and nucleation are relative to each other such that "slow nucleation" implies only that nucleation was not fast enough to create discrete particles before oil separation.

Transition of an oil or an amorphous solid or a crystal can then occur. However, this type of operation can be difficult to control, and scale-up is treacherous because the oil droplets may coalesce into masses and/or form gum balls and increase in size to intolerable levels.

It should also be noted that the tendency to oil out and/or form an amorphous solid is generally increased for low-melting compounds because solvent association can effectively reduce the melting point below its expected value, leading to melting—oiling. High molecular weight compounds are also subject to oiling and/or amorphous solid formation because of the increased complexity of molecular alignment in crystal formation. Oiling is, of course, also dependent on the solubility in the solvent and is particularly likely to occur on "reverse addition" of a solution to an antisolvent. An additional factor may be mutual solubility of the substrate and a component of the solvent mixture, in which case the oil may be a transient solution of the solvent and the substrate in the two-liquid phase, three-component nonequilibrium mixture.

Oiling out may be minimized or eliminated by control of supersaturation and seeding (Deneau and Steele 2005), as discussed below. Seeding has been proven to be essential to prevent oiling out in some systems because, although the oil may not be the thermodynamically stable phase, the transformation to crystals may be sufficiently slow and uncontrolled to cause severe processing problems, as discussed above.

The initial formation of an oil, gel, gum, or amorphous, solid conforms with the Ostwald step rule discussed in Myerson (2001, p. 39). This rule states that in any process, the state that is initially obtained is not the most stable state but rather the least stable state that is closest in terms of free energy change to the original one. It has been postulated that the initial state of crystallization processes is amorphous clusters and that the difference in time constants for the transformation to more stable crystalline states (nuclei) is a key determining factor in the course of the crystallization. It is difficult to distinguish between cluster formation, nucleation, agglomeration, and growth in the early stages (Mersmann 2001, p. 235). The following possibilities can be recognized qualitatively as determined by the time constants and the physical chemistry of the specific compound and system:

- Initial oil or gum that never transforms into a crystal and can agglomerate into large masses.
- Initial oil or gum that transforms into an amorphous solid and stops at that point—never crystallizes—but can form agglomerates.
- Initial oil or gum that transforms into crystals slowly enough that the amorphous form can be observed and can cause agglomeration before discrete crystals are obtained.
- Crystals once formed—either slowly or so rapidly that the amorphous form virtually does not exist—can transform into stable crystals.
- Transformation may continue into more stable polymorphs—if they exist.
- Transformation among polymorphs can be slow such that only the initial form is obtained or, at the other extreme, rapid so that the most stable form is the only one observed.

As is well known, some compounds have never been crystallized, and phase separation results in a stable oil or an amorphous solid. The search for solvents and conditions, or the introduction of foreign particle seeds (e.g., by scratching a glass test tube) to induce crystal formation for a new compound, becomes a matter of trial and error. Combinatorial techniques continue to be developed that can aid in this evaluation. A critical factor for success may be removal of impurities to achieve a very high level of purity, because the effect of even very low levels of impurities on homogeneous nucleation will not be known at this stage.

High supersaturation can lead to very small (nano-sized) particles and resultant agglomeration and gel formation. This phenomenon has been discussed by Mersmann (2001, p. 295) and Mullin (2001, p. 317). Although difficult to define or predict, one mechanism of phase separation leading to agglomeration may be a combination of gelling, to produce a colloidal system, followed by oiling out and nucleation in which the oil serves as a bridge between developing nuclei. Operation at high global or local supersaturation ratios is expected to promote this effect. Mixing can also be a key factor. Although increased mixing may result in breakup of agglomerates in some cases, it may also cause an increased rate of formation by impact between "particles." The effect of mixing on agglomeration of particles smaller than the Kolmogoroff length scale has been termed by Smoluchowski (1918) "orthokinetic." This would predominate in a stirred vessel. The other term used is "perikinetic," pertaining to Brownian motion in a static fluid and when particles are in the submicron size range.

5.4.2 Agglomeration and Aggregation

The distinction between agglomeration and aggregation has been described differently by various authors. Both have received further study by several investigators, including (Myerson, 2001, pp. 110–111, 146), Mullin (2001, pp. 316ff.), and Mersmann (2001, pp. 235ff., 527). In agreement with these authors, the differences between them are not significant and agglomeration will be the term used in this discussion.

One mechanism for agglomerate growth is the collision of growing nuclei followed by "cementing" together from continuing growth between two or more crystals. Although simultaneous collision of more than two particles is not statistically important, the addition of a large number of nuclei to an original two-crystal agglomerate can readily occur by ongoing collisions, leading to very large agglomerates. Aggregation is weak bonding of colloidal particles. Aggregates are relatively easily separated.

Several investigators have developed models for the effectiveness of collisions that lead to agglomeration including Nyvlt et al. (1985) and Söhnel and Garside (1992). This complex interaction of hydrodynamics and crystallization physical chemistry is difficult to predict or describe but can be critical to the successful operation and scale-up of a crystallization process. In particular, for reactive crystallization in which high supersaturation levels are inherently present, agglomeration is very likely to occur as the precipitate forms. Careful control may be necessary to avoid extensive agglomeration, as outlined in Section 5.4.3. below and in Examples 10-1 and 10-2 for reactive crystallization.

The difficulties that can result from agglomeration include

- entrapment of solvent and/or impurities in the crystal mass
- reduced effective surface area for true growth
- subsequent breakup of agglomerates into small crystals that were captured during nucleation without an opportunity for growth
- difficulties in downstream processing because of these small crystals

For these reasons, agglomeration is generally to be avoided. The use of additives (Myerson, 2001, p. 146) may be considered for minimizing agglomeration. However, the use of additives in the pharmaceutical industry—particularly for final products—is generally not done for regulatory reasons barring extreme need.

There are operations, however, which may intentionally generate agglomerates for a particular purpose. These operations are described as flocculation and/or coagulation. However, a discussion of the purposeful generation of these clusters is beyond the scope of this section.

5.4.3 Minimization of Agglomeration

As indicated above, the primary process variables that can be manipulated to minimize agglomeration are

- operation within the metastable region
- controlled rate of supersaturation generation
- removal of crystallization inhibitors from the feed stream
- appropriately high seed levels
- solvent selection
- mixing conditions to achieve particle suspension

These factors are all discussed in the applicable sections of this book and are the same as those recommended for achieving growth.

Both the formation and disruption of agglomerates are functions of mixing conditions and local shear. The reader is referred to the detailed treatment by Mersmann and Braun (Mersmann 2001, pp. 235ff.) in their chapter on "Agglomeration" for a comprehensive analysis of the forces involved in these phenomena. This discussion includes the distinction between attrition and disruption, which are both functions of mixing. Disruption refers to breakup of agglomerates that formed under conditions of low supersaturation that can be broken because the binding forces are small. Agglomerates that are formed under conditions of high supersaturation are much stronger and not subject to disruption by mixing. Attrition refers to breakup of large primary crystals that were formed by growth at low supersaturation and is a function of local shear and crystal lattice energy.

5.5 SEEDING

The influence of seeding on crystallization is often critical to control of a process, and the importance of an appropriate seeding strategy cannot be overemphasized. While some systems will and do nucleate spontaneously, control of a process, especially on scale-up, that depends on spontaneous nucleation can be subject to extreme process variation for several reasons, including

- the presence or absence of seed particles and/or foreign particles from a previous batch
- differences in concentration of impurities from batch to batch that can affect the nucleation rate
- differences in the rate of generation of supersaturation from batch to batch
- differences in the concentration of solute from batch to batch
- differences in mixing scale-up
- local conditions in the crystallizer such as wall temperature, antisolvent addition point, and evaporation rate

The importance of seeding has been underscored by several authors, including Mersmann (2001, pp. 410ff.), Mullin (2001, pp. 197), and Myerson (2001, pp. 240–241) and observed in many processes by the authors of this book, such as Examples 7-1, 7-2, 9-3, 10-1, and others.

5.5.1 Determination of Seed Type, Quantity, and Size

At the conclusion of a crystallization operation, the number and size of the product crystals will be determined primarily by the number and size of the following:

- nuclei generated initially
- nuclei generated during the operation (e.g., by energy input of mixing, local regions of high supersaturation)
- fragments of crystals generated by attrition during the operation
- Ostwald ripening, growth dispersion, and size-dependent growth
- seed added at the outset

Of these critical factors, the one that is most subject to predetermination and control is the seed added at the outset. The following section offers guidelines on effective seeding strategy and methods.

Methods for calculation of an adequate amount of seed are presented below, starting with general guidelines, and are followed by more quantitative approaches.

5.5.1.1 Seeding Guidelines

Four levels of seeding can be described, depending on the purpose of seed addition as follows:

- Pinch, to hopefully avoid oiling out and/or uncontrolled nucleation, crashing, or snowing out. It may be satisfactory in the laboratory but is rarely effective or reliable on scale-up.
- Small ($<1\%$), to hopefully aid in more controlled nucleation but not adequate to achieve primarily growth on scale-up. It is subject to additional nucleation and bimodal distribution of the product.
- Large ($5-10\%$), to improve the probability of growth with the possibility of preventing further nucleation and bimodal distribution.
- Massive (the seed is the product in a continuous or semicontinuous operation), to provide maximum opportunity for all growth. (See Examples 7-6 and 11-6 on resolution of optical isomers.)

The amount of seed can be critical in the control of enantiomer separation and selection between polymorphs and hydrates/solvates. In these applications, nucleation must be prevented to achieve growth of the desired product under conditions in which the undesired isomer/polymorph/hydrate/solvate are supersaturated and therefore could nucleate.

5.5.1.2 Estimation of Seed Types and Quantity

Necessary seed types and quantities to achieve specific amounts of particle size increase in an all-growth process were calculated (Table 4-2) by a simple relationship relating seed and product size to the amount and particle size of seed to be added. Since the validity of this calculation depends on an all-growth process, simultaneous nucleation—which is virtually always present to some degree—will result in a reduction in the actual particle size of the product as well as an increase in the PSD. This increase often comes in the form of a bimodal distribution made up of growth on seed and nucleated smaller crystals. This calculation is useful, however, in predicting the minimum amount of seed required.

It should be noted that the same relationships shown in Table 4-2 for added seed also apply to the number and size of nuclei that are generated without seed. The critical difference between seeds and nuclei is that the amount of seed can be controlled, whereas the number and size of nuclei are difficult to control, as discussed above.

It should also be noted that Table 4-2 indicates that good results in the final mean particle size can be achieved with small amounts of seed. While this may be true algebraically, it is not necessarily true in an actual crystallization because the limited surface area for growth may result in a build-up of supersaturation and nucleation of an excess of particles. Seed plus nuclei may then lead to a smaller final mean particle size and a wide PSD, including the bimodal distribution.

In some cases, seeding may not be necessary because the nucleation rate is very slow and the growth rate is relatively high, thereby allowing growth on a few nuclei without continued excess nucleation. Reliance on this balance of rates on scale-up, however, is extremely risky, primarily because the necessary dependence on spontaneous nucleation to start the process is subject to the batch-to-batch variations discussed above.

5.5.1.3 Seeding Procedures

There are several sources of seed crystals that may be used with varying degrees of effectiveness. Each seeding application requires analysis to determine which of the following sources will be most satisfactory for the operation in question. These sources and methods of preparation include the following:

- Seed from the previous batch—added separately or as heel recycle; effective if the physical and chemical attributes and stability remain satisfactorily constant or can be normalized by intermediate treatment
- Seed prepared in a batch for that purpose and used in many batches; effective provided that the seed so prepared has the necessary physical and chemical attributes and stability
- Seed prepared as above and then milled to achieve a mean particle size and PSD as required; effective for achieving increased control of physical attributes
- Seed from heel recycling with wet milling to control the mean particle size and PSD; one of the most effective seeding procedures when applicable (see Example 10-1)

The issues involved in the necessary record keeping, Good manufacturing practice (GMP) and regulatory control that are associated with a seeded process in the pharmaceutical industry are sometimes cited as reasons not to seed. In response to an article endorsing this premise (Pessler 1997), the following rebuttal by C.B. Rosas (personal communication) summarizes his and the authors' views on this topic.

Not only is timely seeding an excellent way to do away with batch-to-batch vagaries in the width and sensitivity of the metastable, supersaturated region, but the avoidance of nucleation is often crucial for achieving crystallization elegance—a requisite for enhanced rejection of impurities. To trade away such a powerful tool for the sake of trivially lessened inconveniences in record keeping in a GMP plant is most unsound as an operating principle. For those frequent and difficult purification tasks that crystallization from solution does so well, seed early, seed often and, above all, seed always.

5.5.1.4 Seed Preparation

In some cases, a crystallization procedure is unsatisfactory for producing satisfactory seed crystals for subsequent use. This is particularly true for reactive crystallization because of the fine particles normally produced by this method. Since no amount of fine particle seed is satisfactory for significant growth in succeeding batches, the next possibility is to increase the particle size of the seed. To accomplish this, it is necessary to grow seeds in a separate operation from the reactive crystallization. Other applications may also require separate preparation of grown seed. The following procedure is one possible method.

5.5.1.4.1 Growth of Seed—Needles In a separate operation, starting with fine needles, these crystals are subjected to many heat/cool treatments with sonication between cycles. A suitable solvent is required in which the solubility approximately doubles on heating (e.g., a temperature increase of $20-50°C$). During heating, the finer crystals will dissolve. During slow cooling, growth on the remaining crystals may be achieved. Sonication after cooling can break the needles lengthwise, creating more particles. In subsequent cooling cycles, the diameter of the needles, the slowest growth dimension, can slowly increase eventually producing three-dimensional crystals.

This procedure is applicable to other shapes. Crystal breakage by sonication may result in other types of crystal cleavage but three-dimensional growth may also be achieved in these cases.

The needles shown in Fig. 10-3a were subjected to this heat-cool treatment resulting in the large three-dimensional crystals shown in Fig. 10-3b. Success in growing seed crystals both prepares seed for larger batches and establishes that this compound will actually grow. It is then necessary to determine if it will grow at a practical rate in the actual reactive crystallization system.

5.5.2 Effectiveness of Seeding

The effectiveness of seeding is linked to several key factors, including the following:

- timing of seed addition
- condition of the seed surface
- method of seed addition
- rate of generation of supersaturation

5.5.2.1 Seeding Point

The obvious problem of adding the seed before reaching saturation is that of dissolution of some or all of the seed, and the problem of adding the seed after reaching saturation is that of already experiencing nucleation. These difficulties are greatly exaggerated in cooling crystallization of solute with steep solubility dependence on temperature, as described in Example 7-1. In this example, suitable timing of addition of seed was considered to be unachievable in the manufacturing operation and an alternate strategy utilizing seed heel recycling, as described in the example, was utilized.

This problem can also be severe in evaporative methods particularly at reduced pressure, as is common. For these and other reasons described in Chapter 8, evaporative methods are considered to be the most difficult for predictable control of mean particle size and PSD on scale-up.

For antisolvent addition methods, the seed can be added to a small part of the antisolvent, which is then added as slurry as the saturation point is approached. Although some of the seed may dissolve, this technique is considered to be more reliable than adding the seed all at once.

For reactive crystallization, the solubility of the product is usually low and seed can be added before the operation is initiated without concern for dissolution. However, since this method generally produces the smallest particles and is subject to agglomeration, the effectiveness of seeding is critically dependent on providing sufficient surface area for growth at the outset to prevent nucleation at the high local supersaturation conditions at the point of addition and throughout the resulting slurry. The reader is referred to Examples 10-1 and 10-2 for descriptions of operations in which nucleation was minimized in favor of growth.

5.5.2.2 Instrumentation for Seeding Timing

Online instrumentation can be a critical aid in detecting the correct seeding point (Zhou et al. 2006). Measurement of solution concentration by, e.g., Fourier transform infrared (FTIR) near infrared (NIR), UV/visible or Raman spectroscopy can be very effective in determining the seeding point. After seeding, in order to determine whether or not seeding was effective, particle count and size distribution measurement can be made by an in-situ particle size and counting instrument. Sampling and laboratory measurement can be used, but changes that can occur in the sample and the time delay can often introduce unreliability.

Depending on the outcome of the particle number and size determination, corrective measures may be initiated in the event that the seed dissolved or there was excessive nucleation. Reseeding can correct the former and reheating to dissolve excess nuclei may be applicable in the latter.

5.5.2.3 Seed Age

Many crystallization operations can benefit from a seed age in which the temperature is held constant or the antisolvent addition is temporarily halted while the seed crystals grow and deplete the supersaturation before continuing. This procedure can help in normalizing the particle size distribution resulting from the initial steps of seeding and supersaturation generation. The use of online instrumentation can again be very effective in determining when the seed age is complete and the operation can continue.

5.5.2.4 Continuous Operation

The most effective seeding is achieved in semicontinuous and continuous crystallizations by the nature of the operations themselves, in which the seed is always present and in large quantity. Although common in large industrial operations, these techniques have found more limited application in the pharmaceutical industry. Exceptions to this are detailed in Examples 7-6 and 11-6, on the continuous resolution of optical isomers in fluid bed crystallizers, and in Example 10-1, which presents a semicontinuous method of utilizing seed heel recycle in reactive crystallization to achieve primarily growth.

Because they are primarily growth processes, these applications require reduction in seed crystal size to prevent oversized crystals from limiting the surface area. The methods documented in these examples include sonication and wet milling.

5.5.2.5   Condition of Seed Surface

Several authors indicate the importance of the condition of the surface of seed crystals, and some qualitative suggestions can be made. These include (1) making a seed slurry to help condition the surface by dissolution of shards and providing some surface activation (which is also a good method of seed addition) and (2) avoiding the use of dried seed. Milled seed may be necessary for particle size control, but milled seed should either be reslurried or the milling should be done by wet-milling methods.

5.5.2.6   Method of Seed Addition

Seeding with dry powder seed through a vessel head nozzle, while widely practiced in the past, is now limited because of safety and exposure considerations. Slurry seed additions by pump or from seed tanks are preferred and are superior to powder addition for the reasons discussed above. The advantages of retaining the seed in the system, as utilized in continuous operation or in heel recycling are also indicated above.

5.6   RATE OF GENERATION OF SUPERSATURATION

The amount and size of the seed added are critically linked with the rate of generation of supersaturation. The use of increased seed amounts allows an increase in the rate of super-saturation generation by providing more surfaces for growth while avoiding nucleation. Addition and cooling rates are determined by the available surface area as well as by the growth rate of the crystal in question and the width of the metastable region. Seeding is essential, but it is not the only consideration for successful operation.

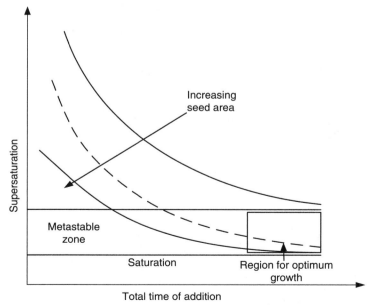

Figure 5-3   Effect of time of addition of an antisolvent or reagent on supersaturation with seed level as the parameter.

These interacting process parameters are illustrated in Fig. 5-3, where supersaturation is plotted against an average solution concentration that would be experienced during addition over the indicated time interval (each point represents an average—one point per run, not a sequence of points in one run). The amount of seed is shown as a parameter in allowing an increased addition rate. This concept is also valid for antisolvent addition time and reactive reagent addition time.

5.7 SUMMARY OF CRITICAL ISSUES

The complex interactions between supersaturation, nucleation, and growth are critical to all crystallization operations. The critical properties of organic compounds that affect these three characteristics include the nucleation rate, growth rate, and width of the metastable zone. These properties vary over large ranges because of the complexity and variety of molecular structures. Crystallization conditions for a particular compound are therefore species dependent and difficult to predict without experimentation.

Nevertheless, some guidelines can be useful, after some critical properties are determined experimentally, and can be used in the development of effective procedures for specific goals. These guidelines may be summarized briefly as follows for processes in which growth is required:

- Seed with crystals of the appropriate amount and size.
- Create and maintain the minimum amount of supersaturation throughout the crystallization.
- Provide effective mixing (see Chapter 6).

For processes for which nucleation is required to produce small particle size:

- Utilize in-line mixing and crystallization techniques to create high supersaturation under controlled conditions (see Chapter 9).

All of the examples in this book are intended to provide information on the utilization of these guidelines in practice.

Chapter **6**

Mixing and Crystallization

6.1 INTRODUCTION

Much of this chapter assumes operation in stirred vessels. Several alternative designs (fluidized bed and impinging jet crystallizers) are summarized early (Table 6-1) for comparison with stirred tanks, and are described later in this chapter and in other parts of this book. An alternative feed addition geometry (mixing elbow) is described in Example 7-1.

Mixing in crystallization involves all elements of transport phenomena: momentum transport, energy transport, and material transport in both the solution phase and the solid phase. In many cases, the interactions of these elements can affect every aspect of a crystallization operation including nucleation, growth, and maintenance of a crystal slurry. To further complicate the problem, mixing optimization for one aspect of an operation may require different parameters than for another aspect even though both requirements must be satisfied simultaneously. In addition, these operations are intrinsically scale dependent.

For these and other reasons to be discussed below, it might be stated that *crystallization is the most difficult of the common unit operations to scale up successfully.*

Successful operations depend on identifying the mixing parameters for the most critical aspects of the process and then evaluating whether those parameters will be satisfactory for the others. Although this approach may be satisfactory in most cases, there will be crystallization procedures that require operation under conditions that are not optimum for mixing for some aspects of the operation, as discussed below.

Note: Successful scale-up implies that both physical and chemical properties have been duplicated between pilot plant and plant operations. These rigid criteria are not always required but are, for example, for final bulk active pharmaceutical ingredients (APIs). In these cases, the width of the particle size distribution (PSD), the average particle size, the bulk density, and the surface area may all be required to fall within specified ranges. It is prudent to apply these criteria, if possible, in bench scale developmental planning and experimentation to reduce the risk of a dramatic failure. Examples of failures that could result from scale-up issues are (1) increased impurity levels, (2) small crystal size causing drastically reduced filtration rates, large PSD including a bimodal distribution, and (3) a poor-washing and slow-drying product.

For final bulk active pharmaceutical compounds, failures could also include physical and chemical properties that do not meet regulatory requirements to meet biobatch specifications.

Crystallization of Organic Compounds: An Industrial Perspective. By H.-H. Tung, E. L. Paul, M. Midler, and J. A. McCauley
Copyright © 2009 John Wiley & Sons, Inc.

Table 6-1   Mixing in Crystallizers for Pharmaceutical Processes

Function	Type of crystallizer		
	Stirred vessel	Fluidized bed	Impinging jet
Continuous and/or batch	Both	Continuous	Semi-continuous
Type of mixer	Variety of impellers	Fluidization	Kinetic energy
Cooling	Good	Excellent	NA
Evaporative	Good	NA	NA
Antisolvent	Good	NA	Excellent
Reactive ppt/cryst	Good	NA	Excellent
Circulation-macromixing	Poor to good*	Excellent	Poor
Mesomixing	Poor to good*	NA	Satisfactory
Micromixing	Poor to good*	NA	Excellent
Micromixing time τ_E, ms	~5 to 40	NA	0.05 to 0.2
Scale-up	Can be difficult	Good with good seed	Excellent
Supersaturation range	Wide	Low	High
Control of supersaturation	Achievable	Excellent at low S	Excellent at high S
Seeding	Wide range	Massive	None or low
Nucleation	Wide range	Minimum	Maximum
Growth	Wide range	Maximum	Minimum

*Dependent on localized conditions in the vessel

6.2   MIXING CONSIDERATIONS

The following is a brief discussion of mixing issues that can be expected to influence crystallization processes. Extended discussions of these and other mixing topics may be found in several references including Baldyga and Bourne (1999), Harnby et al. (1992), and Paul et al. (2003), as well as the crystallization texts of Mersmann (2001), Myerson (2002), Söhnel and Garside (1992), and Mullin (2001). A recent overview of mixing strategies for crystallization is provided by Genck (2003).

As mentioned earlier, mixing requirements for crystallizers involve all aspects of transport properties: momentum, energy, and mass.

For momentum transport, i.e., velocity profile and impact on crystallization, they include

- homogeneity of crystal slurry
- entrainment of gas/vapor from the head space (foaming)
- secondary nucleation through impact
- shear damage to crystals and impact on agglomerate formation/breakup
- satisfactory discharge of the slurry without excess retention of product crystals and possible operation over a wide volume range

For energy transport, i.e., temperature profile and impact on crystallization, they include

- rate of heat transfer through the jacket wall
- avoidance of encrustation—solid scale on walls and baffles

For mass transport, i.e., solution concentration profile and impact on crystallization, they include

- blending of solution and antisolvent components to the molecular level to achieve uniform supersaturation
- blending of reagents to the molecular level to achieve reactive crystallization/precipitation

For example, the initial blending of components to the molecular level, while avoiding regions of high supersaturation, requires consideration of the mesomixing and micromixing environments of the contactor. Other requirements involve the macromixing capabilities of the crystallizer. Extensive discussions of these fundamentals of mixing may be found in the references cited above.

Although many variations of mixing systems have been used for crystallization processes, the three primary types that are discussed in this book are the stirred vessel, fluidized bed, and impinging jet devices. Each of these utilizes different mixing environments to achieve the desired local and global conditions.

The predominant system in the pharmaceutical industry is the stirred vessel. Fluidized beds (Chapters 7 and 11) and impinging jets (Chapter 9) fill specific mixing requirements, as indicated in Table 6-1.

6.3 MIXING EFFECTS ON NUCLEATION

6.3.1 Primary Nucleation

The effects of mixing on primary nucleation are exceedingly complex. The overall result is a reduction in the width of the metastable region when this width for a static solution is compared to that for an agitated solution (see Chapter 2). Therefore, an unagitated solution can, in general, be cooled further before the onset of nucleation than an agitated solution. Since an industrial system with few exceptions will always be agitated, this is of theoretical interest only. (Exceptions are for operations such as melt and freeze crystallization in which such issues are key factors; see Chapter 11 in this book; Mullin 2001, pp. 343ff.; Mersmann 2001, pp. 663ff.)

In a mixed solution without crystals present and at constant supersaturation, increased mixing intensity can reduce the induction time—the time elapsed after mixing to create supersaturation to the time crystals first appear. Induction time decreases up to a critical speed, after which it remains unchanged (Myerson 2001, p. 145). Additional discussion may be found in Chapter 4.

6.3.2 Secondary Nucleation

Since secondary nucleation is dominant as soon as nuclei appear, the nucleation mechanisms become virtually impossible to characterize in an industrial operation. In addition, any seeded crystallization is by definition secondary even though some nuclei may simultaneously form by a primary or other secondary mode mechanism. Therefore, the majority of this discussion will focus on secondary nucleation.

The effect of agitation on secondary nucleation has been reported in the literature, and several references are discussed by Mullin (2001). Secondary nucleation is mixing dependent as follows:

- Crystal-crystal impact: a function of both the local micromixing environment and the overall macromixing circulation
- Crystal-impeller and crystal-wall impact: functions of impeller speed, shape of the blade, and material of construction

These factors, along with the other intrinsic nucleation properties of the crystallizing substrate, affect the rate of nucleation, which in turn determines the number of nuclei formed and their size. This complex relationship is strongly dependent on the specific system characteristics but, in general, the nucleation rate increases rapidly with increasing energy input. This high dependence is especially true for reactive crystallization with fast reactions. The reader is referred to Chapter 10 for a discussion of this topic, as well as to Söhnel and Garside (1992).

6.3.3   Damkoehler Number for Nucleation

It is helpful to visualize the relationship between mixing and nucleation rates through an analogy with the reaction Damkoehler number (Da). The Da number for reaction is defined as

$$\text{Da for reaction} = \text{mixing time/reaction time}$$

in which mixing time and reaction time can be the time required from the initial state (time zero) to 95% or other percentages of the final state. The mixing time is usually a measure of local mixing where the reaction is occurring. However, local mixing times vary over a wide range in a stirred vessel in which the reactants are subject to varying local mixing times.

As shown in Fig. 6-1, at low values of the Da number, the reaction yield is insensitive to mixing, whereas at high values of the Da number, the reaction yield is sensitive to mixing. The reader can find a more detailed discussion on this concept in chapter 13 of Paul et al. (2003).

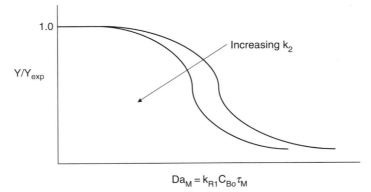

Figure 6-1   Reaction yield as a function of the reaction Da number; mixing sensitivity for chemical reactions is analogous to mixing sensitivity in crystallization nucleation and growth, from low sensitivity at low Da to high sensitivity at high Da.

By analogy, a Da number for nucleation could be visualized as a ratio between mixing time and induction time:

$$\text{Da for nucleation} = \text{mixing time/induction time}$$

Mixing time and induction time can be the time required from the initial state to 95% or other percentages of the final state.

As shown in Fig. 6-2, at low values of this ratio—fast mixing and long induction time—mixing would have a minimal effect on nuclei size, whereas at high ratios—slow mixing and short induction time—mixing effects would be critical to nuclei size. Local supersaturation could be visualized as a parameter as shown.

For slow-nucleating systems that have long induction times or equivalently wide metastable zone widths, the location of the feed point and the impeller energy input may not be as important as for higher values of the ratio in determining the size—and number—of nuclei that can be generated in regions of the feed stream. This analogy also parallels that used in reacting systems in evaluating the requirements for micromixing of reagents.

For reactive crystallization processes, in addition to mixing time and induction time, we may need to consider the reaction time in the analysis. A fast reaction (high Da for reaction) could be readily seen as sensitive to the local mixing environment of two reactive streams. It could generate locally a high product supersaturation due to the fast reaction. This, in turn, could generate small particles as in precipitation which can have very short induction time. However, if the product has a long induction time, i.e., low Da for nucleation, the local high supersaturation regions can be mixed with the rest of the system and distributed throughout the vessel before nucleation. The PSD would not be sensitive to micromixing or mesomixing but may still be sensitive to macromixing and other mixing issues, such as impact with the impeller and other crystals.

If the reaction rate is slow compared to the mixing rate, the reactants can be dispersed throughout the vessel before they react. The resulting product solution concentration will be more uniform throughout the vessel. The operation could be thought of as being determined by crystallization parameters rather than the reaction rate.

For pharmaceuticals, most of the reactive crystallization processes are salt formation from an acid and base. In this situation, the reaction rate is generally much faster than the mixing or crystallization rate, and the mixing sensitivity depends primarily upon the magnitude of the induction time, or Da for nucleation.

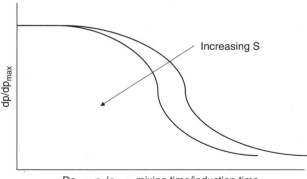

Figure 6-2   Particle size as a function of a crystallization Da number (the ratio of mixing time/nucleation induction time).

It should always be kept in mind that mixing and nucleation time can be arbitrarily defined as the time required from the initial state to, for example, 95% of the final state. These values are also affected by other operating factors, such as the type of solvent, temperature, degree of supersaturation, flow ratios of two streams, and vessel geometry.

It cannot be overemphasized that the Da number also depends upon the "scale." On the one hand, we can see the mixing time as primarily determined by the scale of the mixing device. Mixing time in micro-scale devices, which can be considered equivalent to the time for micro-scale mixing, is always faster than the mixing time in large-scale vessels. On the other hand, induction time is not considered to be a function of scale. Thus, a nucleation process may perform successfully in the laboratory using a small vessel (mixing time shorter than induction time) but fail to produce the expected mean particle size and PSD in the pilot plant or factory vessels (mixing time longer than induction time) because of the local non-uniformity of concentration and its effect on nucleation induction time.

Clearly, the key to a successful scale-up of this process is to maintain the mixing time at the pilot plant and factory scale below the threshold which would cause the process to move down the curve of Fig. 6-2. This can be best accomplished by using a special mixing device, such as an impinging jet, to approach the same mixing time at all scales. The reader can find more information on impinging jets and crystallization in Section 6.6.3 and Examples 9-5 and 9-6.

6.3.4   Scale-up of Nucleation-Based Processes

Nucleation events can dominate the entire crystallization operation with respect to both physical and chemical purity attributes. Ultimate crystal size as a function of the number of nuclei generated is summarized in Table 4-2, where the nominal dimensions of resulting crystals (spherical, flat plates, needle-shaped) are shown as a function of the number of nuclei or seed particles added. It can be seen that the number of nuclei generated by the various causes of nucleation—including agitation—has a negative exponential effect, as expected from this purely geometrical relationship, on the ultimate size that can be achieved by growth subsequent to nucleation. The nucleation rate, particularly contact nucleation, can increase on scale-up because all key parameters of (stirred tank) mixing cannot be held constant. For example, scaling up at equal power per unit volume results in an increase in impeller tip speed. The resulting average particle size could then be reduced because, after the increased nucleation, there are more particles for a reduced amount of substrate to grow on. In addition, other mixing factors that affect growth could increase the size distribution further, as discussed in Section 6.4. Oiling out and agglomeration also can further complicate a nucleation-based process, as discussed in Section 5.4.

The critical nature of these interactions is the key factor in causing difficulty in scale-up of nucleation-based crystallization processes—even with small quantities of seed.

The critical mixing factors in a stirred tank are impeller speed and type, as well as their influence on local turbulence and overall circulation. Since all aspects of these factors cannot be maintained constant on scale-up either locally or globally, the extent to which changes in the crystallizing environment will affect nucleation is extremely difficult to predict. To the mixing issue must be added the uncertainties caused by soluble and insoluble impurities that may be present in sufficiently different concentrations from batch to batch to cause variation in induction time, nucleation rate, and particle size.

The severe problems associated with nucleation-based operations, some of which, especially in stirred vessels, are directly caused by mixing issues, lead to the conclusion that dependence on nucleation can be problematic in achieving reproducible results on scale-up and/or in ongoing production.

If no process alternative is possible to avoid dependence on nucleation, mixing scale-up can be based on equal power per unit volume, assuming that the same impeller type is used. In most cases, however, this approach will result in changes in PSD on scale-up that may or may not be acceptable. In general, as suggested by Nyvlt (1971), the PSD will be broader and the average particle size will be smaller if this scale-up criterion is used. As often experienced in crystallization scale-up, however, the opposite can occur, depending on the specific nucleation characteristics of the system. A further generalization is that rapid nucleation tends to produce the smaller size distribution on scale-up, whereas slow nucleation can give the opposite result.

A more robust way to accomplish consistent scale-up of a nucleation-based process is to apply impinging jet crystallization. The energy of impinging jets can achieve rapid mixing of two streams at high local energy dissipation rates to the molecular level, possibly before nucleation occurs. By matching the energy dissipation rate at different scales, the desired mixing time from laboratory-scale devices to production scale can be achieved. This technology is discussed in Chapters 9 and 10.

Several references on nucleation provide excellent insight into this complex phenomenon, including Mersmann (2001), Mullin (2001), and Myerson (2001).

6.4 MIXING EFFECTS ON CRYSTAL GROWTH

6.4.1 Mass Transfer Rate

Mixing can obviously have a large effect on the mass transfer rate of growing crystals through its effect on the film thickness. This influence is dependent on both the size of the crystals and the mixing intensity. As mixing intensity increases, mass transfer rate increases and film thickness decreases up to a limit, beyond which the effects approach limiting values.

The concentration gradient in the film is illustrated in Fig. 6-3, where it can be seen that supersaturation conditions are present throughout the film. When crystallization is diffusion limited, concentration in solution drops significantly from that in the bulk (C_b) to that in the film (C_f), which is close to that at the crystal growth surface (C_{surf}). When growth rate is primarily limited by resistances to surface incorporation, the larger drop in concentration is

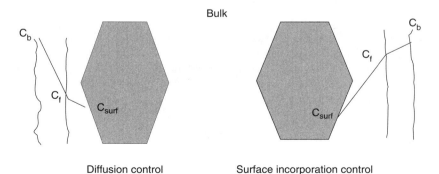

Bulk

Diffusion control   Surface incorporation control

Figure 6-3   Schematic representation of concentration gradients from bulk solution to growing surface.

across the film. The thickness of the film will determine the time available for nucleation within it. The new nuclei are either shed into the bulk or attach as shards to growing surfaces. In both cases, the effect will be to limit growth.

Large crystals (>100 microns) may be more subject to film mixing issues than smaller ones because crystals less than 10–20 microns are approximating the Kolmogoroff eddy size and tend to follow these eddies. These differences in crystal size can change the growth rate limitation from film control in large crystals to intrinsic growth rate control in small ones.

6.4.2 Da Number for Crystallization

The Da number concept has been applied to crystallization by Garside (1971) and Garside and Tavare (1985) by the development of an effectiveness factor for growth by comparing the growth rate at the interface conditions to the growth rate expected if the interface were exposed to the bulk solution conditions. The Da number in their work is defined as the ratio of the surface integration rate to the mass transfer rate through the film. This definition is different from the Da number defined in this book, which is the ratio of mixing time to crystal growth or nucleation time. The effectiveness factor varies in an S-curve from 1.0 at low Da to zero at high Da. A value of 1.0 would indicate pure surface integration growth where mixing effects are unimportant, while a low value approaching 0 would indicate pure diffusion growth with high sensitivity to mixing. Regions between these limits represent a combination where both growth and mixing have an effect.

The Da number for crystal growth can be defined as

$$\text{Da for crystallization} = \text{mixing time/overall crystal growth time}$$

or equivalently

$$\text{Da} = \text{mixing time/supersaturation release time}$$

Again, mixing time and supersaturation release time can be those required to achieve 95% or percentages of the final state.

Similar to the discussion of the Da number for nucleation, at low Da values for crystallization—fast mixing and slow release of supersaturation—mixing would not affect crystal growth, whereas at high ratios—slow mixing and fast release of supersaturation—mixing effects would be critical to PSD, as fast local crystal growth may occur.

For systems with a slow release of supersaturation, the location of the feed point and the impeller energy input may not be as important as for those with higher values of the ratio in determining the size of crystals that can grow in regions of the feed stream.

Depending upon the nature of the crystals and the operating environment, authors have observed a wide range of release rates of supersaturation. The time scale for the release of supersaturation can vary from fractions of seconds to days (or longer). Due to the wide variation in release rate of supersaturation, the inherent concepts of the Da number for crystallization provide a useful way to readily identify the potential mixing sensitivity issue of a particular crystallization process.

6.4.3 Conflicting Mixing Effects

Factors that improve with increased mixing in a stirred tank are (1) heat transfer, (2) bulk turnover, (3) dispersion of an additive such as an antisolvent or a reagent, (4) uniformity of crystal suspension, (5) avoidance of settling and minimization of wall scale, and (6) minimization of impurity concentration at the crystallizing surface.

However, these factors must be balanced against the possibly negative results of overmixing that can result in crystal breakage and/or shedding of nuclei as well as increased secondary nucleation.

These concerns lead to the conclusion, referred to above, that it is often necessary to choose a mixing condition (impeller speed, type, etc.) that may not be optimum for every aspect of the crystallization and may actually not be optimum for any of them. In many cases, however, one end result (i.e., PSD, bulk density, uniformity of suspension, and approach to equilibrium solubility [yield]) may dictate the choice of mixing conditions. In this case, it becomes essential to determine if the negatively affected aspects can be tolerated. If these problems are occurring in operation in a stirred vessel, a different type of crystallizer, such as a fluidized bed, might be used to promote crystal growth and minimize nucleation. Readers can find more information on fluidized bed crystallizers in Section 6.6.2 and Examples 7-6 and 11-6.

6.4.4 Experimentation on Mixing Effects

All of these factors are properties of a given crystallization system, thereby requiring choices for each specific operation. Experimentation is required to determine the key responses to mixing for each system and could include determination of the following:

- Effect of impeller speed and type on PSD at a minimum of two seed levels and two supersaturation ratios. These results should indicate the sensitivity of the system to mixing.

Note: A small response could indicate that other system properties were controlling (i.e., inherent crystal growth rate or nucleation rate). A large response would indicate sensitivity to secondary nucleation and/or crystal cleavage and require additional experimentation and evaluation of scale-up requirements. The laboratory results should be evaluated relative to each other since scale-up can be expected to make additional changes in PSD, especially when a large response is experienced in these simple experiments.

- Effect of impeller speed on crystallization rate and approach to equilibrium solubility (yield). Failure to achieve equilibrium solubility may indicate accumulation of impurities at the crystallizing surfaces. An increase in impeller speed resulting in further reduction in solution concentration could indicate resumption of growth or additional nucleation (see Example 6-1).
- Suspension requirements, as indicated by the settling rate to achieve off-bottom suspension (see Section 6.6 below).
- Effect of feed pipe location (for antisolvent and reactive crystallizations) (see Section 6.6.1.5 below).

For nucleation-dependent operations, it is recommended that additional information be obtained as follows:

- Effect of impeller speed and type on the width of the metastable region.
- Effect of impeller speed and type on the rate of nucleation.

Note: This experimentation is focused primarily on evaluation of mixing sensitivity. Other experimentation on crystallization issues is beyond the scope of this discussion.

6.4.5   Effects of Mixing on PSD

The effect of mixing on PSD has been experimentally examined for the reactive crystallization of calcium oxalate (Marcant and David 1991). This work is an excellent example of the multiple dependencies on mixing that can be experienced in a crystallization operation. The factors noted above that are mixing dependent are shown to have positive or negative influences on the resulting physical characteristics, thereby illustrating the necessity of selecting the most important result to be achieved. Increasing the agitator speed is shown to initially cause an increase in particle size, followed by passing through a maximum and then decreasing particle size. This result is attributed to changes in controlling factors resulting from the changes in mixing. This experimental result provides an excellent justification for both variable-speed drive and subsurface feed.

Experimental results indicating the sensitivity of particle size and PSD to the location of the feed stream may be used to confirm the sensitivity of a compound to mixing. If no sensitivity is observed, it may be concluded that intrinsic factors such slow growth or low nucleation rate are dominant. For additional discussion, see Section 6.6.1.5 below.

6.5   MIXING SCALE-UP

6.5.1   Power

The compromises in mixing optimization that may be required on scale-up often result in the use of the common mixing criterion of equal power per unit volume or, in some cases, equal tip speed. Both of these recommendations are more relevant for utilization of the same impeller type as well as geometric similarity. Laboratory evaluation of the mixing requirements for a crystallization operation should be carried out in a minimum 0.004 m^3 liter vessel (4 liters) for preliminary data and a further evaluation at 0.1 to 1 m^3 as practical.

Smaller-scale operations will generally produce a more uniform PSD and a larger mean crystal size than the manufacturing scale (typically $\sim 10\,m^3$) when using equal power per unit volume. These changes typically are caused by the local differences in impeller shear (an unavoidable result of the equal power per unit volume criterion) that cause increased nucleation leading to a larger number of particles, an increased spread in PSD, and a smaller particle average diameter.

Guidelines other than equal power per unit volume were suggested by Nienow (1976) that can be helpful in avoiding this local over-mixing. Using this guideline, the agitator speed at the manufacturing scale would be selected to be sufficient to just maintain off-bottom suspension, thereby resulting, in addition to reduced shear damage, in reduced nucleation, fewer particles, and more growth. In general, this speed would be considerably less than equal power per unit volume, depending primarily on the density difference between the suspending solvent and the crystals and their size.

Limitations on this guideline would be high-density crystals that require possibly damaging higher speeds to achieve the just-suspended condition. In addition, antisolvent and reactive crystallization applications may require higher speeds to prevent local supersaturation at the point of addition. In the latter case, scale-up based on equal local energy dissipation at the point of addition may be necessary.

A further caution on reduced speed is a possible increase in encrustation caused by crystal contact with the bottom surface with insufficient fluid force to prevent wall growth.

6.5.2   Off-Bottom Suspension

The requirements for particle off-bottom suspension are also discussed in many literature references including Zwieterling (1958), Chowdhury et al. (1995), and Paul et al. (2003, Chapter 10). Methods of calculating various degrees of homogeneity in solids are presented. These calculations are an important part of crystallization scale-up studies.

6.6   CRYSTALLIZATION EQUIPMENT

The three types of crystallizers summarized in Table 6-1 achieve operation in extreme mixing environments, thereby providing a wide variety of contacting capabilities. Equipment choice can be tailored to specific needs of supersaturation control—high or low—and mixing intensity—high or low—as well as other aspects of the operation. However, as discussed in the sections that follow, it may be difficult to find one mixing system for a stirred tank that can satisfy all the needs of a specific crystallization operation. Fluidized beds and impinging jets can be designed and operated in narrower regions of mixing intensity, low and high, respectively, and may therefore be tailored to meet all the needs for a particular operation when matched to the specific needs of that system.

6.6.1   Stirred Vessels

6.6.1.1   Alloys and Stainless Steels

A stirred vessel crystallizer is shown in Fig. 6-4. Included are a dual-impeller pitched-blade turbine with a "tickler" blade (see Section 6.6.1.6), a subsurface addition line, baffles, and a ram-type bottom outlet valve to aid in discharge of slurries.

The workhorse impeller is the pitched-blade turbine because of its ability to create good circulation at relatively low shear. These attributes help reduce secondary nucleation and crystal breakage while achieving good suspension and circulation. The flat-blade turbine is less versatile because of high shear and less overall circulation. The Ekato Intermig has proven to have superior performance in some crystallization operations because of its combination of excellent circulation and low shear.

Baffles are required in all cases to prevent poor mixing due to swirling as well as entrainment of vapor which can provide nucleation sites. Baffles may also have an important effect in minimizing foaming. Propellers and hydrofoils are not normally suitable for multipurpose service in heterogeneous systems—especially in the event that nucleation and growth may pass though a stage involving sticky solids, as discussed above in Section 5.4.1.

Computational fluid dynamics representations of flow in these vessels may be helpful in visualizing flow patterns and particle paths.

6.6.1.2   Glass-Lined Vessels

The versatility of the glass-lined vessel in a large variety of chemical environments has made it the most common in the industry. For these reasons, a crystallization step may be carried out in an equipment train in a glass-lined vessel whether or not this is required to prevent corrosion. These reactors range in size from 80 to 20,000 liters or more. One limitation in the use of glass-lined vessels related to mixing and heat transfer is that the limit of the temperature difference between jacket and batch is ~125°C. However, such temperature extremes

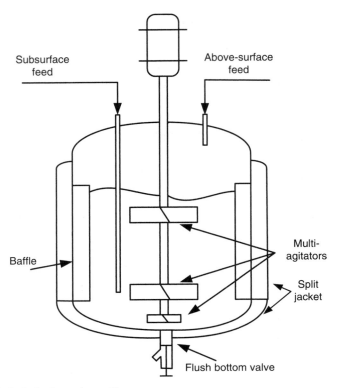

Figure 6-4 Typical stirred vessel crystallizer.

are not often encountered in crystallization. (The manufacturer should be consulted for specific limitations on the type of glass lining and base metal in use.)

The retreat-blade and anchor impellers that have been widely used for many years are now being replaced by glass-lined turbines and other shapes that have been recently developed by the manufacturers using sophisticated methods of applying the glass to more sharply angled shapes. These turbines are available for vessels as small as 80 liters, although shafts with removable, interchangeable impellers are not currently available for tank sizes smaller than about 1200 liters. These limitations are subject to additional improvement, and it is recommended that the specific vendors be consulted for updated information.

These new impellers, especially in multitier configurations, have greatly improved the mixing capabilities of the glass-lined reactor by providing increased shear and circulation. A number of glassed impeller types are now available including curved- and pitched-blade turbines. Some examples of these are shown in Fig. 6-5. In addition, two or three turbines mounted on a single shaft are now available for larger vessels ($>3\,m^3$). The lower turbine can be positioned within $\sim 10\,cm$ of the vessel bottom. For single turbines in larger vessels, however, the low turbine position may not provide the desired overall circulation.

The three manufacturers of glass-lined mixing systems use different methods of attachment of the blades of the impeller(s) to the shaft as summarized in Table 6-2.

A single agitator adapted for each requirement

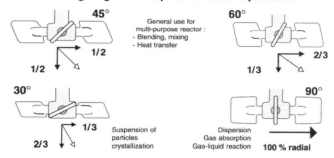

Figure 6-5 (a) GlasLock® glass-lined steel impellers.
Courtesy of De Dietrich Process Systems.

6.6.1.3 Baffles

Although glassed baffle design has significantly improved mixing performance, most glass-lined vessel applications utilize only a single baffle in order to maximize the number of tank nozzles available for other purposes. Two manufacturers, DeDietrich and Tycon, offer modifications to include multiple baffles in their mixing systems.

6.6.1.4 Variable-Speed Drive

The use of variable-speed drives in pilot plant and manufacturing plant vessels is recommended for development and scale-up of crystallization processes. This capability provides the opportunity for critical experimentation at the pilot plant scale to determine the effect of impeller speed on PSD and other variables. On the manufacturing scale, the ability to change impeller speed is the most readily adjustable parameter for manipulation on scale-up. Modern variable-frequency drives provide an excellent means to vary speed over a wide range. The added cost of variable-speed capability is minimal compared to all other methods of changing mixing

(b)

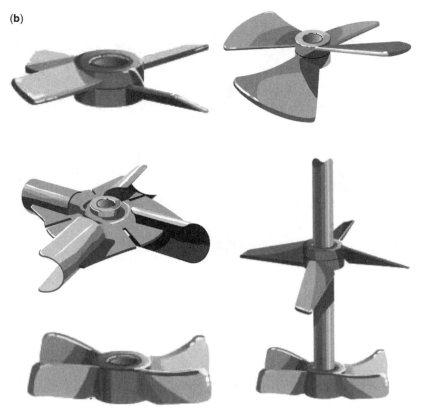

Figure 6-5 (b) Cryo-Lock® impellers.
Courtesy of Pfaudler, Inc.

parameters and may be one of the most effective ways of solving scale-up problems and providing versatility for multipurpose operations. In addition, speed changes can be utilized in troubleshooting in plant operations that are not meeting expected results.

6.6.1.5 Subsurface Addition Lines

Another key factor in successful pilot plant and manufacturing plant operation is the appropriate use of subsurface addition of antisolvents and reactive reagents for crystallization and precipitation. The primary purpose of subsurface addition is the introduction of the agent causing supersaturation in the region of intense micromixing in the vessel to avoid local excess supersaturation with the associated potential for increased nucleation.

The effect of the feed point has been shown to be dramatic in the reactive crystallization of calcium oxalate, as reported by Marcant and David (1991). The conclusion was reached that determination of the sensitivity of crystallization to feed point location can provide information on controlling factors. In a laboratory development program, therefore, an effective method of determining the importance of mixing effects is to run experiments with two or more feed points—a surface feed and an impeller feed—and at two different impeller speeds. If no difference can be found in the PSD and mean particle size, mixing effects may not be controlling and other aspects such as inherent growth rate may be dominant. As discussed in

Figure 6-5 (c) ElcoLock® and fixed impellers.® impellers.
Courtesy of Tycon Technologies, a Robbins & Myers Company.

Sections 6.3.1.4 and 6.4 on the use of the Damkoehler number concept, the need for effective location of a feed line may be evaluated from a comparison of nucleation and mixing rates.

The mixing texts referenced above contain extensive discussion of the importance of the location of feed streams. While these studies are primarily concerned with reagent feed for chemical reactions and the influence of local turbulence on reaction selectivity, the same issues are encountered in the addition of antisolvents and reagents for reactive crystallization because nucleation is a function of supersaturation, whether local or global.

The location of the point of introduction and the diameter and flow rate of the addition stream are key to successful operation and scale-up. Generally, the optimum location is at the

Table 6-2 Glass-Lined Impellers and Their Methods of Attachment

	Blade attachment	Removable	Variable pitch
DeDietrich	GlasLock® Friction fit of blades into holes in the hub	Yes	Yes (Within power limit of motor and drive)
Pfaudler	Cryo-Lock® Liquid N_2 cooling to shrink the shaft	Yes	No (Different pitches available)
Tycon	Elcolock®	Yes	No (Different pitches available)

point of maximum turbulence, as characterized by the shortest mixing time constant. This point is in or near the impeller discharge flow. For the down-pumping pitched-blade impeller, this point would be just above the impeller so that the flow would be rapidly mixed by passing through the impeller blades. For a radial flow flat-blade turbine, it would be directed into the discharge flow.

Feed introduction on or near the surface of a stirred vessel can have dramatic effects on a crystallization process because unwanted oiling out, nucleation, and/or agglomeration can occur in these poorly mixed zones. It is recognized that mechanical design and cleaning issues may make the use of subsurface lines more difficult. However, it cannot be overemphasized that the negative effects on a manufacturing-scale operation can be far more costly than provision of the necessary mechanical requirements. For alloy vessels, the mechanical issues are minimal. For glass-lined vessels, the use of Teflon tubing attached to the baffle has been effective.

Cleaning issues can be resolved by providing for removal of the subsurface line between batches or during turnaround.

The diameter of the subsurface line is also a key factor with crystallization, as it is with reactions. If the diameter is too large, the incoming stream may produce regions of high supersaturation before it can be effectively blended to the molecular level. Large-diameter pipes can also be subject to reverse flow in which the crystallizing mixture is forced into the pipe, where it is subject to meeting the incoming antisolvent in a nearly stagnant region, leading to nucleation and possible solids accumulation and plugging.

Extensive analysis and experimentation on dip-pipe design has been carried out by Penney and co-workers (Jo et al. 1994), and the reader is referred to this work for specific design recommendations for diameter and flow rate. Their recommendations are summarized in Table 6-3 for six-blade flat-blade and three-blade high-efficiency down-pumping impellers. These recommendations should be followed to prevent dip-pipe back-mixing, enhance local blending, and prevent solids plugging.

In the Table 6-3, D and T are the impeller and tank diameter, respectively, G is the distance from the feed pipe to the impeller, and $V_f/V_{t,\min}$ is the recommended minimum ratio of the velocity in the feed pipe to the velocity in the tank at the feed pipe location.

The importance of the feed location for chemical reactions has been clearly established by the work referenced above and many others. The literature contains less data on crystallization. However, undesired nucleation is potentially present for all crystallization systems, depending on the nucleation rate and the degree of local supersaturation. The analogy between reaction sensitivity and supersaturation sensitivity can be visualized through the concepts represented by the Damkoehler number, as discussed in Section 6.3.1.4 above.

Table 6-3 Recommended Minimum V_f/V_t for Selected Geometries for Turbulent Feed Pipe Low Conditions

Case	Impeller	Feed position	D/T	G/D	$V_f/V_{t,min}$
1	6BD*	Radial/mid-plane[‡]	0.53	0.1	1.9
2	6BD*	Above/near shaft[§]	0.53	0.55	0.25
3	HE-3[†]	Radial/mid-plane[†]	0.53	0.1	0.1
4	HE-3[†]	Above/near-shaft[§]	0.53	0.55	0.15

*Six-blade disk turbine.

[†]High-efficiency three-blade down-pumping turbine.

[‡]Injection radially inward toward the impeller at its mid-plane at a distance G.

[§]Injection downward into the impeller at about $D/4$ from the centerline of the impeller shaft and G/D above the impeller mid-plane.

As in scale-up of chemical reactions, it is important to consider the relative time constants of mixing and nucleation in order to estimate which part of the S-curve applies to a particular system. Although this determination is conceptual, the insight provided by such analysis can be helpful in establishing operational design criteria.

6.6.1.6 Discharge of Slurry

Efficient discharge of the crystal slurry to the next process vessel—usually a filter—is required to prevent yield loss by settling or to achieve the necessary degree of crystal slurry homogeneity for satisfactory filter loading or other subsequent operations. The mixing obviously changes as the slurry is discharged and can result in an unsatisfactory degree of settling after passing the impeller or an unsatisfactory degree of homogeneity for the subsequent operation. "Tickler" impellers are often used, as shown in Fig. 6-4, to provide mixing until the discharge is nearly complete. These may be custom designed and fitted by the manufacturer. The diameter of the tickler blade is usually one-half to two-thirds the diameter of the main impeller, and thereby requires little power and does not create sufficient shear to be harmful.

Selection of pumps for discharge may also be a significant issue with regard to energy input and slurry pumping capability. Crystals that have been carefully grown and are ready for high rate filtration can be reduced in size by pump shear to drastically reduce the filtration rate. This is especially problematic for recycle pumps in filter equipment feed loops.

Low-shear pumps can be effective in limiting shear damage and include lobe and diaphragm pumps. The reader is referred to an analysis of pump and transfer energy by Mersmann (2001, pp. 454ff.) to aid in design of slurry transfer and control of flow.

Shear damage on transfer can be eliminated by gravity transfer when equipment layout includes the necessary vertical clearances.

6.6.2 Fluidized Bed Crystallizer

An important alternative type of crystallizer is the fluidized bed. One fluidized bed design is shown in Fig. 6-6. As shown, the supersaturated solution enters at the bottom of the crystallizer and the clear, partly or fully depleted solution exits at the top of the crystallizer. The slurry is suspended in the crystallizer by the upward liquid flow, which usually lies in the

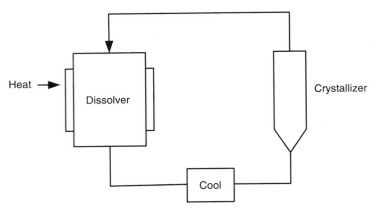

Figure 6-6 Fluidized bed crystallizer and dissolver.

laminar and transition flow domain. When designed with a tapered bottom section of modest included angle to minimize flow separation, the hydrodynamic flow pattern is much closer to the plug flow within a tube than that in a continuously stirred tank. Therefore, the crystallizer can exhibit a high degree of homogeneity with a very low degree of back-mixing at the same height of the bed, but the slurry concentration and particle size distribution can vary along the height of the bed. This is a sharp contrast to the "homogeneous" mixing of slurry within a stirred tank.

Due to its low turbulence intensity, fluidized bed crystallization is very effective in minimizing nucleation by providing very low shear, low energy, and minimum velocity impact between crystals. The operation can achieve growth by operating at low supersaturation and in the presence of a large surface area for growth.

This technology is used primarily in cooling crystallization instead of antisolvent or reactive crystallization. Thus, residence time, instead of mixing time of two streams, would be more appropriate for the definition of the Da number, as well as for scale-up design. A major advantage of continuous and semicontinuous operation that can be realized with fluidized beds is that the seed is always present, thereby eliminating the need for timing of seed introduction and other aspects that have been discussed as variables in batch and semibatch operation. Initial seed preparation becomes critical to success; methods for growing initial seed are presented in Chapter 5. Once grown, however, the ongoing growth on the seed continually renews the supply. Methods of crystal cleavage to maintain seed PSD are also presented in Chapter 5.

Both continuous and semicontinuous operation can be utilized. This principle has been successfully applied in the resolution of optical isomers in which nucleation must be minimized, and preferably eliminated, to achieve isomer separation, as described by Midler (1970, 1975, 1976) and presented in Examples 7-6 and 11-6 in this book. Tools for monitoring PSD online, as mentioned in Section 2.10.2, are very applicable here.

6.6.3 Impinging Jet Crystallizer

Impinging jet crystallization achieves the opposite extreme of mixing, compared with fluidized bed crystallization, of high shear and energy input in small regions. As shown in Fig. 6-7, two (or multiple) streams of high velocity are impinged upon each other, which results in a rapid localized intense mixing of these streams. Depending upon the mixing

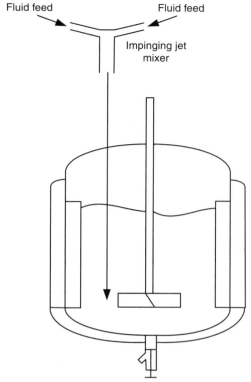

Fluid feed   Fluid feed

Impinging jet
mixer

Figure 6-7   Impinging jet crystallizer.

intensity and detailed design of the impinging jet device, the mixing time of the two streams can be fractions of a second and can be shorter than the induction time. As a result, this device can be used to generate a high level of supersaturation for a nucleation-dominant process. At the optimum mixing effectiveness, an impinging jet can mix the feed streams to the molecular level in less than the induction time, thereby allowing nucleation under uniform conditions throughout the operation.

This principle has been successfully utilized in an industrial application to achieve a small average particle size (3–5 microns) and a narrow PSD. For impinging jet crystallization, industrial operation is described by Midler et al. (1994), with variants by Lindrud et al. (2001) (impinging jet crystallization with sonication) and by Am Ende et al. (2003) (specific reference to reactive crystallization). Laboratory studies are reported by Mahajan and Kirwan (1996), Benet et al. (1999), Condon (2001), and Hacherl (Condon) (2003). Johnson (2003) and Johnson and Prud'homme (2003) report on the use of impinging jets to produce nanoparticles stabilized by block copolymers.

EXAMPLE 6-1

Process: Crystallization of an intermediate (MW ~ 700) from a very impure mixture.

Issues: Slow growth rate, shear damage during long approach to equilibrium. Loss of yield because actual equilibrium solubility was not achieved.

Probable Cause(s)

- Accumulation of the impurity(ies) at the crystallizing surfaces, limiting the growth rate.
- Thick films around the crystals because of limited impeller speed to prevent crystal damage.

Possible Solutions

- Extend the age time (provides more time for the solute to reach the crystal surface).
- Remove impurities (reduces impurity accumulation at the surface and thereby speeds up the growth rate).
- Increase impeller speed (reduce film thickness) or reduce it (less shear damage).

Laboratory evaluation indicated that the expected equilibrium solubility could be achieved through longer aging (in the laboratory). However, the plant operation did not improve with longer aging.

Removal of impurities at this stage was not feasible. The intermediate produced in this step is recrystallized to improve purity, and good growth is achieved in the purified system. Therefore, it is known that the compound will grow.

SOLUTION: It was observed that adequate growth occurred early in the process and then stopped. To test the ability of fluid forces to reduce the presumed impurity film on the crystal surfaces, impeller speed was increased. An increase in impeller speed by ∼20% after growth had stopped was effective in reaching the equilibrium and achieving the expected yield.

Conclusions

An increase in impeller speed was effective either because (1) the impurity film was reduced, allowing growth to resume, or (2) the increased shear caused additional nucleation.

Message

Impurity(ies) can have a profound effect on the growth rate and, in this case, on the approach to equilibrium solubility. As in many cases due to experimental time constraints, the actual cause of the improvement was not clearly established. A very effective method of determining the impact of impurity(ies) from the process in question is to recrystallize the compound from a pure solvent(s) and from its own mother liquor and compare crystal size, growth rate, and morphology.

Incorporation of impurities in the crystals was a major concern that indicated higher impeller speeds. However, excessive fines and the resulting poor filtration rates were counterbalancing influences on the determination of impeller speed. The two-level agitation rate scheme was a balance between these conflicting factors that resulted in passable purity, yield, and filtration rate. As in many high-impurity systems, however, the average particle size was small because of the need to operate at relatively high supersaturation to achieve practical growth rates, thereby incurring more nucleation than desired.

The qualitative aspects of this difficult crystallization provide examples of the conflicting requirements that are often encountered in development and scale-up. ∎

Chapter 7

Cooling Crystallization

Crystallization by cooling is commonly practiced for solutions in which solubility is a strong function of temperature. Cooling alone can achieve the desired degree of crystallization when solubility is sufficiently low at the termination of the cooling operation. In some cases, additional reduction in solubility is necessary to achieve the desired yield. This reduction can be accomplished by either evaporation or antisolvent addition, as discussed in Chapters 8 and 9.

Crystallization by cooling can be carried out in a batch or continuous operation. The reader is referred to Chapter 1 for a discussion of the advantages of each. In this chapter, a description of crystallization by batch cooling will be followed by a description of the continuous operation. As discussed in the introduction, a semibatch operation is one carried out with changes taking place (in volume, composition, etc.) which affect the critical parameters. An operation, or part of an operation, is considered semicontinuous if, during the operation, conditions in the processing unit being considered are unchanging.

7.1 BATCH OPERATION

A typical cooling operation, as described by Griffiths (1925), is shown in Fig. 7-1. Solubility versus temperature is shown by A-B and the width of the metastable region is bounded by A-B and C-D. A solution at point E which is below the equilibrium saturation solubility is cooled until the temperature reaches the equilibrium solubility at E′, at which point crystallization could start. In most cases—depending on a variety of factors including the rate of temperature reduction, ability of the compound to sustain supersaturation, the presence or absence of seed, and the presence of impurities—crystallization will not start at this point but will require some degree of supersaturation before being initiated. This could occur within the metastable region, point F, in which case the concentration could follow along F-G. On the other hand, crystallization may not begin until the temperature reaches point H and could then follow H-G.

Curves F-G and H-G represent two different modes of carrying out crystallization by cooling and would be expected to result in two very different results in terms of physical properties—mean particle size (mean dp) particle size distribution (PSD), surface area, and bulk density, as well as possible differences in rejection of impurities and occlusion of solvent. Other differences can include a change of morphology and a potential for changes in polymorph formation. These differences result from the supersaturation and surface

Crystallization of Organic Compounds: An Industrial Perspective. By H.-H. Tung, E. L. Paul, M. Midler, and J. A. McCauley
Copyright © 2009 John Wiley & Sons, Inc.

area histories during the cooling operation, which in turn control the degree of nucleation and growth.

Nucleation and the distinctions between primary and secondary nucleation, and homogeneous versus heterogeneous nucleation, are discussed in Chapter 4. As noted in that chapter, more detailed discussion of these topics may be found in the excellent treatments of Myerson (2001), Mullin (2001), and Mersmann (2001). In the current discussion, the different types of nucleation are not separated because they often occur simultaneously and can result from more than one of these mechanisms.

For crystallization by cooling, the following factors can affect the supersaturation profile either locally or globally:

- rate of cooling and wall temperature
- width of the metastable region
- nucleation rate and inherent crystal growth rate
- presence or absence of seed and seed quantity
- mixing and mass transfer
- solvent system
- impurities (dissolved and undissolved)

A high slope of solubility versus temperature may require tailored process conditions to avoid localized regions of excessive supersaturation, which can, in turn, cause undesired nucleation and fines in the product. See Example 7-1, which illustrates one solution to this problem.

7.1.1   Rate of Cooling

The rate of cooling is critical in several ways. Referring again to Fig. 7-1, the rate of temperature change as the saturation curve at point E′ is approached will influence how far into the

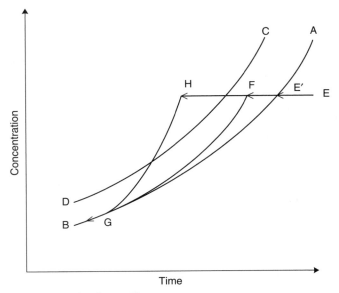

Figure 7-1   Solution concentration time profile.

metastable region the temperature will be before nucleation sets in. In the case of rapid cooling, the temperature can reach point H before nucleation is induced at this high degree of supersaturation.

Figure 7-2 shows the well-known relationship between nucleation rate and supersaturation and illustrates that the number of nuclei that can be generated by a high cooling rate increases exponentially with increasing supersaturation. *This number of nuclei, in turn, influences the rest of the crystallization operation* for the following reasons:

- A high number at the outset limits growth potential because the ultimate size of the particles is determined by their number. See the discussion of this issue in Section 5.2.1.

- A high degree of supersaturation can also cause oiling out and/or enhance agglomeration/aggregation of the developing nuclei, as well as cause occlusion of impurities and/or solvent in the crystals (see Section 5.4).

- Agglomeration and aggregation are difficult to predict. Control and scale-up of those processes that pass through stages of agglomeration should be avoided or minimized if possible by the cooling strategies discussed below.

- The rate of cooling is one key variable. Applying a fixed-temperature coolant to the jacket or cooling coils results in "natural" cooling, as illustrated in Fig. 7-3. This results in a high rate of cooling with the resulting risk of high supersaturation and nucleation. The other consequence of natural cooling can be exaggerated wall growth resulting from local high supersaturation at the cooling surface.

- Linear cooling is one means of reducing the initial cooling rate, but ideally, temperature-cooling strategies can be utilized to match the cooling rate with the increasing surface area. These rates were derived by Mullin and Nyvlt (1971) and Nyvlt et al. (1973). Data-based cooling curves for maximization of final particle size were derived by Jones and Mullin (1974) and further refined by Jones (1974). Wey and Karpinski [in Myerson 2002, pp. 244–245] provide a good discussion of the current status of the effort in this area, which has been very useful for industrial practitioners.

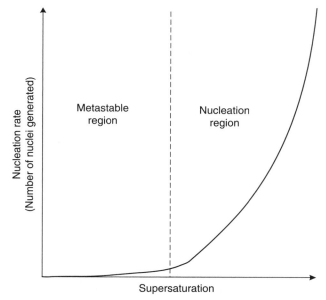

Figure 7-2 Nucleation versus supersaturation.

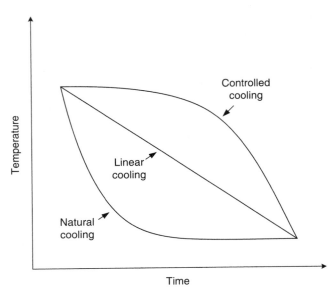

Figure 7-3   Natural cooling versus linear and controlled cooling.

As shown in Fig. 7-3, the prescribed cooling rates are much slower at the outset than the natural cooling rate. This is necessary to maintain supersaturation in or close to the growth region when the surface area for growth is low. The cooling rate can be increased as the surface area increases. An added benefit of this method is the potential to reduce scaling by limiting temperature differences across the jacket.

In initiating a cooling operation, the cooling rate should be reduced as the saturation temperature is approached. After seeding or initial nucleation, the temperature can be held constant for a period of time to allow the seed to develop, as discussed by Jones and Mullin (1974) prior to controlled cooling for the remainder of the cycle.

7.1.2 Metastable Region

The lower concentration limit of the metastable region (A-B in Fig. 7-1) is the equilibrium solubility of the compound in the solvent system. The upper limit (C-D) represents a supersaturation ratio below which nucleation is relatively slow but above which nucleation will be spontaneous. Chapters 1 and 2 give more detailed discussion on the metastable region.

Unlike the lower limit, the upper limit is not a result of thermodynamic equilibrium because it is a function of kinetic factors such as rate of cooling. However, it provides a very useful guide in the design of a crystallization process. The width of this region, in terms of the temperature difference between the limits, is a qualitative indication of the ability of the compound to remain supersaturated when the concentration exceeds saturation. Extreme differences have been observed for this temperature range for different compounds or for the same compound at different temperatures, in different solvent systems, and with different levels of dissolved impurities.

In general, to minimize nucleation and achieve growth, it is desirable to avoid operation above the upper limit, C-D. The possibility of maximizing growth can be realized by operating as close to the saturation curve, A-B, as possible while still maintaining a supersaturation ratio that will give a satisfactory growth rate. This applies to both local and global supersaturation ratios. Methods of achieving this growth regime are discussed later in this chapter and specifically in Example 7-1.

Another factor in the crystallization of organic compounds is the tendency of some to oil out. This tendency increases rapidly in the nucleation regime and provides a further incentive to control the supersaturation with appropriate cooling strategies. The reader is referred to Section 5.4 for a discussion of this phenomenon.

7.1.3 Seeding versus Spontaneous Nucleation

As noted in previous chapters, the influence of seeding on crystallization is often critical to control of a process, and its importance cannot be overemphasized. Many or most organic solutes in crystallizations driven by cooling will nucleate spontaneously if the cooling is sufficiently rapid. However, control of a process, especially on scale-up, that depends *primarily* on spontaneous nucleation can be subject to extreme process variation.

Several factors can be responsible for this unpredictability, including the following:

- Cooling rates are generally reduced at larger scale, primarily because of lower heat transfer surface-to-volume ratios.

- For those systems which maintain the same cooling rate on scale-up, there are generally greater localized differences in supersaturation within the larger vessel because of higher wall film temperature gradients (a lower-temperature coolant may be utilized or required on the large scale).

- There are differences in bulk mixing which are inherent on scale-up.

- There is possible batch-to-batch variation in dissolved and/or undissolved impurities, which can influence the nucleation rate.

These issues can cause large differences in the number of nuclei generated, which in turn can cause differences in final PSD and other chemical and physical attributes, as discussed above. Therefore, scale-up can fail both because the laboratory or pilot scale results cannot be achieved on the large scale or simply because of severe batch-to-batch variation. Both can affect downstream processing or, in the case of a final product, the physical attributes that may be critical to the end use of the product.

A typical seeded batch cooling crystallization is illustrated in Example 7-2. Also, see Example 7-3 for another instance of seeding from a different perspective.

7.1.4 Mixing and Mass Transfer

Mixing may affect crystallization by cooling in all of the ways outlined in Chapter 6. In particular, wall film thickness and the resultant heat transfer rate are most affected, along with local supersaturation profiles at the cooling surface.

7.1.5 Solvent

Crystallization by cooling is normally carried out in a single solvent—one of the advantages of this method of generating supersaturation. In many cases, however, the substrate may be in a solvent mixture to start out, or a second solvent may be added to reduce the ultimate solubility to increase the yield.

The choice of solvent depends primarily on the dependence of solubility on the temperature that determines both the maximum concentration that can be achieved at the initial temperature and the yield at the final temperature. In addition, the solvent can have a very significant effect on several of the factors discussed above, including width of the metastable region, nucleation, growth rate, and crystal shape. These effects must be determined experimentally, although generalizations can be made regarding some physical and chemical factors, such as hydrogen bonding capability, polarity, and dielectric constant. In addition, the ability of the solvent to form solvates and to affect the crystal form and morphology can be critical. The science of these complex interactions is addressed in Chapter 2 and comprehensively discussed in several texts, including Mullin (2001) and the chapter of Meenan et al. in Myerson (2002).

7.1.6 Impurities (Dissolved and Undissolved)

The importance of the effects of dissolved impurities on nucleation and growth rates cannot be overemphasized. They are extremely system dependent, since the magnitude of their influence depends primarily on molecular structure similarities and dissimilarities between the substrate and the impurities. These structural parameters determine, in turn, the amounts (concentrations) of impurities that will exert significant effects. In most cases, therefore, some experimental data must be obtained on the specific system being considered.

It is important to recognize the importance of these factors in the design of a crystallization process. If significant effects are observed or inferred, control of upstream processing to maintain consistent feed is critical.

Figure 7-4 illustrates the effects of impurities on the formation of different crystal form of one drug substance. For this particular compound, the data show that conversion from the less stable form to a more stable form is much slower or undetected in the actual mother liquor that contains dissolved impurities (upper curve) than it is in pure solvent (lower curve).

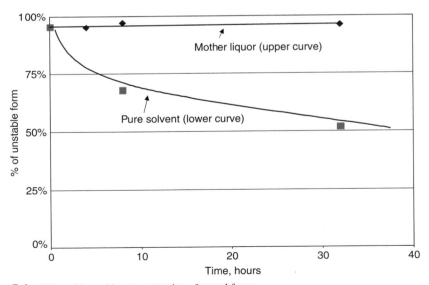

Figure 7-4   Effect of impurities on conversion of crystal forms.

Undissolved impurities can influence a process primarily by acting as seed. In some cases, this can be beneficial if seeding is not being used as part of the process. However, dependence on foreign particles as seed is generally not satisfactory.

7.2 CONTINUOUS OPERATIONS

The difference between batch, semibatch, continuous, and semicontinuous processing was discussed in Chapter 1. Continuous processes are characterized by parameters which may be geographically distributed within the system but are unchanging with time. Continuous crystallizers are in common use throughout the chemical process industries, but are less so in the pharmaceutical and fine chemical industries because of the typically smaller amounts to be processed. This section on continuous cooling crystallization will discuss continuous operation and point out the differences from batch/semibatch operation described above. It will also illustrate some strategies and equipment types used for these operations.

7.2.1 The Attraction of Continuous Processing

The chemical industry has long understood the advantages of continuous processing for many unit operations, including crystallization. Firstly, control of critical parameters in a process is much more easily obtained in time-invariant, steady-state operation than in trying to optimize an entire batch trajectory. Secondly, continuous operation ensures, at least in theory, that every molecule in a container of product was treated the same way as every other. In practice, differences caused by residence time distribution effects may significantly influence product quality attributes and must be taken into account in the design of these operations (Paul et al. 2003, chapter 1). Variations due to time and global/local conditions that are inherent in batch/semibatch operation are also likely to be significant in continuous operation. Since continuous crystallizers do not require the time-consuming (and sometimes difficult) development of a new seed bed for each batch in a growth crystallization process, they are generally able to produce the desired amount of material in smaller equipment. This can result in significant capital savings on both the equipment and its footprint space in the factory.

The above advantages of continuous over batch operation can often be lost in the pharmaceutical industry, however, for the following reasons:

- Product requirements are often too small to justify the startup and shutdown effort and material required to establish steady-state operation.
- Batch operation permits harvesting at equilibrium (saturation).
- Maintenance of batch integrity simplifies the quality control function.
- Batch operation allows frequent cleaning of the equipment without upset to the system.

Having noted above the merits of batch processing in some circumstances, it is important to remind the reader that there are crystallizations for which batch operation cannot achieve the desired separation in a practical process. Examples 7-4 and 7-6 illustrate this important point.

7.2.2 Operating Strategy for Continuous Cooling Crystallizers

Figure 7-5 shows a typical flow pattern for a continuous or semicontinuous cooling crystallizer. The system might or might not have a "fines trap," which can take many forms (a cyclone, for example) but is most often a heated redissolution section somewhere within the apparatus. The entering feed stream and the product outlet stream are usually continuous rather than intermittent for better control. Intermittent product removal is sometimes used, however, for less critical control of slurry flow.

A generalized schematic diagram for feedforward/feedback control of particle size in a continuous or semicontinuous crystallizer, such as that in Fig. 7-5, is shown in Fig. 7-6. The measurement device can be most easily envisioned as an in-line particle size analyzer such as a Lasentec focused beam reflectance measurement (FBRM), but it could also be a sampler/offline device combination (even a plain microscope). The most common variant on this control schematic would be elimination of feedforward control, and perhaps of the fines trap as well.

In Fig. 7-6, m_i is the ith moment of the particle size distribution ($i = 0$ [number], 1 [diameter or length], 2 [area], or 3 [mass or volume]), whichever makes the most sense from the point of view of the particular process being carried out.

Figure 7-7 illustrates three common flow patterns for operation of a mixed suspension crystallizer. In each, M is the magma (slurry) outlet and O is the overflow.

The first flow pattern in Fig. 7-7(a), shows the usual mixed suspension, mixed product removal (MSMPR) pattern. The most common example of this is a well-mixed continuous stirred tank reactor (CSTR).

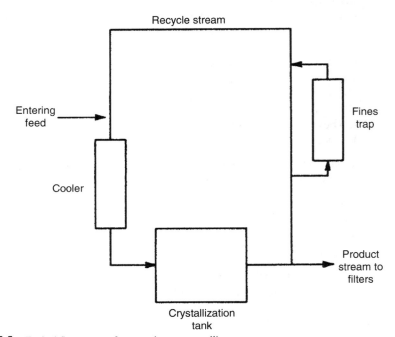

Figure 7-5 Typical flow pattern for a continuous crystallizer.

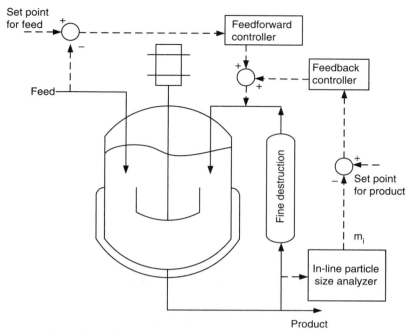

Figure 7-6 Feedforward/feedback crystallizer control.

The second, (b), shows clear liquid overflow, as in a well-behaved fluidized bed crystallizer or in the semicontinuous stirred tank (SCST) operation illustrated in Example 7-5. The SCST has proven to be a versatile and practical method of maintaining control in critical separation and polymorph control, and can be operated in small and large configurations.

The third, (c), is clear liquid overflow recycling to an elutriating leg which prevents undersized particles from escaping prematurely. Draft tube baffle (DTB) crystallizers contain elutriation legs, but DTB units are more commonly evaporative rather than cooling crystallizers.

Figure 7-8 shows a simplified information flow diagram for a continuous MSMPR crystallizer. Population balance equations (see Chapter 4) can be used to separate nucleation and growth effects. For particles keeping geometric similarity, the surface area of the particles for

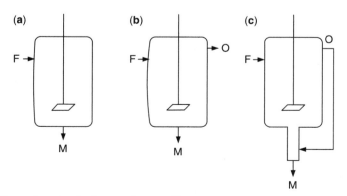

Figure 7-7 Flow patterns in mixed suspension crystallizers.

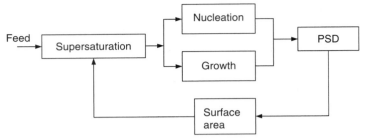

Figure 7-8   Simplified information flow for an MSMPR crystallizer.

a given resident mass is reduced linearly with increased linear dimension of the crystals. This reduced surface area is less capable of removing supersaturation in the crystallizer, so the supersaturation increases. Operating at steady state, a well-run continuous crystallizer maintains these factors in balance while producing product crystals with satisfactory particle size distribution.

7.2.3   Plug Flow and Cascade Operation

In order to increase the driving force for crystallization or increase the yield per pass through the system, a continuous crystallization system can be intentionally operated as a cascade, as shown in Fig. 7-9. A significant number of these crystallizers in series become, in effect, a plug flow reactor.

Fluidized bed crystallizers operated in particulate mode (usually the case with liquids) can also come very close to plug flow operation. An example of such crystallizers will be shown in Example 7-6.

7.2.4   Fluidized Bed Continuous Cooling Crystallizer Designs

The pure fluidized bed crystallizers described later in Example 7-6, while designed for a particular purpose, have many similarities with the Oslo commercial surface-cooled crystallizer shown in Fig. 7-10. A fluidized magma in the crystallizer body (E) carries out the crystallization. Feed enters the clear overflow stream (at G), is cooled within the metastable supersaturation region in the cooler (H), and then enters as a fluidizing stream at the bottom of (E).

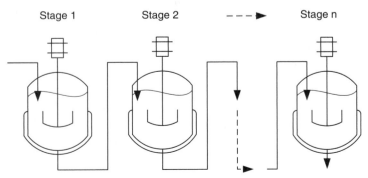

Figure 7-9   Cascade operation.

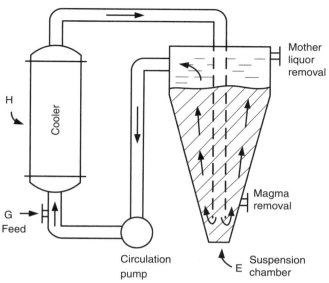

Figure 7-10 Flow in an Oslo cooling crystallizer.

Fluidized bed designs have a number of advantages, including the capability for fines removal from the main crystallizer by elutriation (carrying out with the fluidizing liquid) and low shear mixing by fluidization without the need for mechanical agitation.

Resolution of optical isomers via preferential crystallization is outlined in Example 7-6 as an example of the use of tightly controlled supersaturation in a cooling crystallization. This process is discussed in greater detail in Example 11-6.

Example 7-6 illustrates the applicability of good crystallization practice to achieve continuous production of large-volume pharmaceutical compounds. It also illustrates a crystallization process that is inherently unfeasible by any method other than continuous operation. When carried out using fluidized bed crystallizers, ultrasonic crystal disruption is used, even at factory scale, to maintain a steady-state population of seed particles in this all-growth system.

7.3 PROCESS DESIGN—EXAMPLES

Design of a crystallization process requires consideration of the impact of many of the above factors on the resultant chemical purification and physical properties. Process intermediates and final bulk products often require very different crystallization operations in order to achieve specific process outcomes. The following examples illustrate both types of process issues and solution as well as different process designs to achieve specific physical and/or chemical properties.

> **EXAMPLE 7-1** *Intermediate in a Multistep Synthesis*

> *Goal*: Maximize growth to prevent excessive fines and occlusion of impurities

> *Issues*: High slope of temperature versus solubility curve
>   High nucleation rate results in wide PSD
>   Seed point is difficult to define

Crystallization of the penultimate intermediate in a multistep synthesis was required primarily for separation of the intermediate from accumulated organic impurities. The previous reaction and subsequent extraction are both run in isopropyl acetate (IPAC) and are fed to the crystallization step at a temperature ($80°C$) that is slightly above the saturation temperature ($\sim 70°C$) in the water-saturated IPAC solution.

Crystallization is readily induced by seeding and cooling to $60-65°C$. However, the crystals (Fig. 7-11) obtained by natural cooling or by programmed ("cubic") cooling are too small for the subsequent filtration operation with regard to time of filtration, occlusion of impurities, and drying (high solvent content of the wet filter cake).

This common problem is particularly acute in this operation because of the very high slope of the solubility curve for the compound in water-saturated IPAC, as shown in Fig. 7-12. The high slope causes two different problems, both of which require control to achieve crystal growth and impurity rejection to achieve the process goals cited above. The issues are (1) when to seed so that the seed does not dissolve (temperature too high) or nucleation occurs before seeding (temperature of seeding too low), and (2) how to control supersaturation during cooling.

Seed addition either too soon or too late results in uncontrolled nucleation, which in this case produces excessive fines and reduced potential for growth. The issue of when to seed is often critical, and the actual saturation temperature for optimum seed addition is subject to some batch-to-batch variations because of possible changes in concentration and/or impurities from the previous reaction(s) and purification(s).

Control of supersaturation is difficult in cooling operations when the solubility slope is high and small temperature changes result in creation of supersaturation ratios outside the metastable region either globally or locally (i.e., at the cooling surfaces).

These issues were experienced in pilot scale operation and would be expected to be acute on scale-up to production.

In many cooling crystallization operations, these issues do not present a serious problem either because the metastable region is wide, the nucleation rate is low, or the compound has high growth potential. However, in this example, special measures were needed to achieve the required process outcomes.

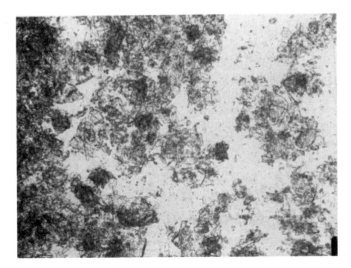

Figure 7-11 Product from seeding and cooling process ($200\times$).

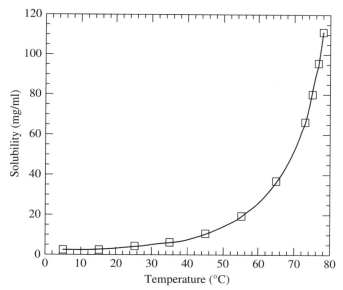

Figure 7-12 Steep temperature-solubility slope for a desired intermediate.

SOLUTION: Constant Temperature, Semibatch Addition of Feed, High Level of Seeding, In-line Mixing All attempts to run this crystallization by conventional cooling failed to produce the crystal growth needed for a satisfactory filtration rate, impurity rejection, and drying rate. These problems were overcome by development of a process utilizing semibatch addition of the IPAC feed solution to a recycling seed bed through an in-line mixing elbow followed by cooling of the crystallization batch. Flow in the mixing elbow is shown in Fig. 7-13.

This design utilizes an impinging jet feed mixing entry, which achieves compositional uniformity (eliminating localized zones of high supersaturation) at a time scale smaller than that required for nucleation. It incorporates a swept elbow through which the vessel slurry is recycled by a low-shear pump. The hot batch is then introduced in a jet-like manner as an opposing flow into the recycling heel.

Both the feed stream and the circulating slurry stream should possess enough energy to create good micromixing. Ideally, the feed stream should have a linear velocity in the neighborhood of 5 m/sec to achieve this purpose, but in fact the mixing elbow has been found to be effective at somewhat lower feed velocities. The recirculating slurry velocity should be at least slightly lower than that of the feed stream to avoid the possibility of backmixing into the feed pipe. However, to avoid compromising the desired macro-, meso-, and micromixing provided in this device, the slurry should be circulated at a velocity of at least about 1 m/sec. Computational fluid dynamics (CFD) simulation (not experimentally validated) indicates that effective compositional uniformity is achieved in several hundred milliseconds in this system.

An IPAC solution of the feed coming from a continuous extractor is fed through the mixing elbow to a recycling slurry of seed and crystallizing product at a temperature of 65°C over an extended time period. During this time, the volume of the slurry in the crystallizer increases three-fold as all of the feed is charged. The resulting slurry is then cooled to 0°C, aged, and filtered.

The operation is initiated by recycling, through the mixing elbow, ~15% of the previous batch slurry as seed.

This procedure addresses the issues outlined above as follows:

- The large amount of seed presents a high-growth area from the outset of the operation. The initial presence of the seed avoids the issue of when to add it to prevent seed dissolution and homogeneous nucleation. Note that for this process, the size of the particles does *not* need to decrease, as steady-state particle size distribution is reached at the low but finite natural attrition rate achieved in the pump-around and agitation systems.

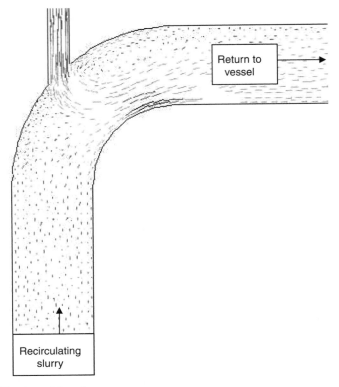

Figure 7-13   Flow in a mixing elbow.

- Rapid addition avoids nucleation from local supersaturation by mixing prior to formation of critical-sized clusters.

- Adiabatic addition avoids possible nucleation from heat transfer surfaces.

- Approximately 50% of the product crystallizes during the addition at 65°C. The mixing tee combines the incoming feed and the recycling batch sufficiently rapidly to prevent local supersaturation levels that would cause excessive nucleation. The slow addition rate combined with the high growth area prevents supersaturation levels that would cause nucleation by allowing the crystal growth rate to be limiting.

- Cooling after completion of the semibatch addition can proceed without concern for nucleation because ~50% of the product is present as crystals, thereby providing a large growth area and avoiding excessive supersaturation.

Pictures of the crystals obtained by this procedure are shown in Fig. 7-14. Filtration rates and impurity rejection are satisfactory. Scale-up from the pilot plant to manufacturing scale (10,000 liters) resulted in no significant differences in crystallizer performance.   ■

EXAMPLE 7-2   *Pure Crystallization of an API*

Goals: To understand crystallization behavior upon seeding, aging, and cool-down

Variables: Seed level, nucleation/growth at seeding temperature and cool-down profile

Figure 7-14   Product crystals (200×).

In this example, the final pure bulk pharmaceutical recrystallization procedure is designed to purify crude and generate the PSD that meets all pure bulk physical property requirements without use of milling. This is accomplished by employing a crystallization procedure that ensures a growth-dominant strategy, followed by vacuum drying, delumping, and drying. Occasional PSD reproducibility problems in laboratory and pilot plant performance motivated an in-depth investigation of several aspects of the pure recrystallization procedure.

In this procedure, the batch is heated to achieve complete dissolution. After cooling the batch to reach a slightly supersaturated state, a finite amount of seed is charged. The seeded batch is aged and then cooled down slowly to its final temperature for harvesting. The target parameters are seeding temperature, seed level, seeding/aging temperature, aging period, initial cool-down temperature, and cool-down period. A bifactorial experimental design was used to define the range and number of experiments. Lasentec FBRM was utilized to obtain in-situ particle size distribution (PSD) information. A Microtrac laser diffraction analyzer was used to obtain offline PSD information of the filtered dry cake sample. Sizes noted in the description below are all area mean calculations.

TRANSIENT PSD RESPONSE

Figures 7-15 and 7-16 plot the transient responses of total particle counts and mean particle size after seeding and during the aging period. When seed is charged to a supersaturated solution (Curve I of Fig. 7-15), the total particle count increases rapidly and then levels off during aging. Concurrently, mean particle size (Curve I of Fig. 7-16) decreases quickly after seeding and then levels off during aging. This behavior indicates the occurrence of rapid nucleation after seeding in a supersaturated solution, followed by crystal growth on the seed crystals and newly formed nuclei. It is noted that immediately after seeding, the mean particle size shows a fast oscillatory response. This is presumably due to the dispersion of seed agglomerates and nucleation on the seed crystals simultaneously.

When the solution is saturated or undersaturated, an interesting opposite response is observed. The total particle count decreases (Curve II of Fig. 7-15), and the mean particle size increases (Curve II of Fig. 7-16). This behavior strongly implies digestion of particles, especially fines of the (milled) seed.

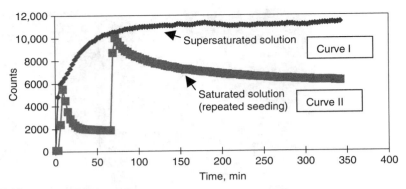

Figure 7-15   Profiles of total particle counts.

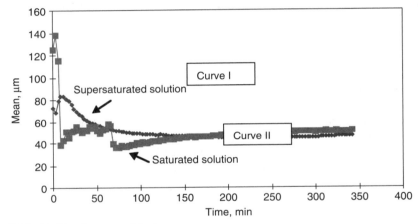

Figure 7-16   Profiles of mean particle size.

Particle count and mean size during cool-down are shown in Figs. 7-17 and 7-18. After the start of cool-down (~200 minutes), the total particle count increases only slightly, whereas the mean particle size increases noticeably. These observations suggest that crystal growth is dominant in this period.

FINAL MEAN PARTICLE SIZE

A total of six parameters—seed level, seeding temperature, seed aging period, initial cool-down period, second cool-down temperature, and third cool-down period—were evaluated. In studying the effect of these six parameters, the best-fitted models (by least squares) of the data are:

Lasentec (in-line): Mean (μm) $= 62 - 19.5 *$ (seed level in %) $+ 0.026 *$ (initial cool-down

period in minutes)

Microtrac (off-line): Mean (μm) $= 61.4 - 24 *$ (seed level in %) $+ 5.4 *$ (seeding temperature in °C)

As indicated by both measuring devices, seed level is determined to be the most critical parameter affecting the final mean particle size. Within the range of seed level for this study (1 $\pm$ 0.5 wt%), with a 0.5 wt% increase of seed, the mean particle size will decrease by 10 μm (0.5 * 19.5), as predicted by the Lasentec model, and by 12 μm (0.5 * 24), as predicted

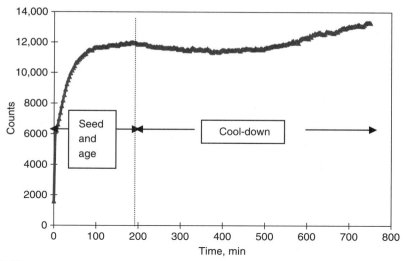

Figure 7-17 Profile of total particle counts during cool-down.

by the Microtrac model. This finding emphasizes the critical role of seed in the development of the crystallization process.

The above equations show that of the remaining parameters, seeding temperature and length of the initial cool-down period were most significant. This reemphasizes the importance of control of supersaturation in this process.

FACTORY SCALE-UP

After identifying the critical parameter through the laboratory investigation, the recrystallization procedure was scaled up successfully from the laboratory (500 ml scale) to the factory (1100 liter scale). Figure 7-19 compares the final PSD of laboratory and factory materials.

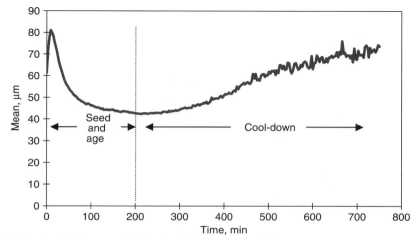

Figure 7-18 Profile of mean particle size (area-based) during cool-down.

> **EXAMPLE 7-3** *Crystallization Using the Heel from the Previous Batch as Seed*

Goals: Consistent control of crystal surface area and crystal shape without charging dry seed from batch to batch

Variables: Seed level, seed surface area, and particle breakage

This crystallization procedure ensures a growth-dominant strategy in order to consistently reach the desired purity and final PSD. The goal of this example was to control the final crystal surface area and shape in order to optimize the dissolution rate of the compound.

This batch cooling procedure is very similar to the recrystallization procedure employed in the previous example. In some experiments, sonication is applied at the end of crystallization and a small portion of the slurry is retained as seed for the next batch. Seed level, seed surface area, and sonication were varied in the investigation.

CRYSTAL SURFACE AREA AND SHAPE

Figure 7-20 shows the microscopic photos of final crystals under the same magnification at 1 wt% and 5 wt% seed level, respectively. These results clearly demonstrate the impact of seed level on final crystal size.

To illustrate the impact of seed surface area and sonication, Table 7-1 summarizes the results of four consecutive crystallization experiments at the 3 wt% seed level. In each experiment, the seed used is the final crystals from the previous experiment. As can be seen from the table for the first two experiments, seed surface area has a strong impact on the final product surface area. This is anticipated when crystal growth on the seed bed is the predominant mechanism.

Because the availability of the growing surface of seed will be reduced from batch to batch as the seed grows, eventually the available growth surface area will be so low that it cannot relieve the supersaturation in the allotted time cycle. Under this circumstance, significant nucleation will occur, and many smaller particles will be generated. The fresh small crystals will contain an abundant growing surface area, and the process will repeat again as a growth-dominant process until it reaches the next cycle. In order to overcome this undesired oscillatory behavior of the PSD and the surface area, sonication was used to break up the crystals and to provide fresh surface area for crystal growth. As shown in Table 7-1 for the third and fourth experiments, with sonication the product surface area becomes steadier and is much less sensitive to the seed surface area.

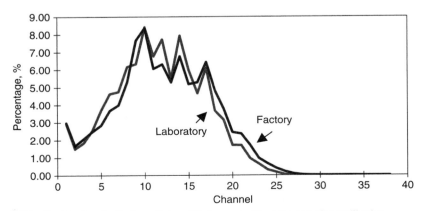

Figure 7-19 Particle size distribution of lab and factory materials at the end of crystallization.

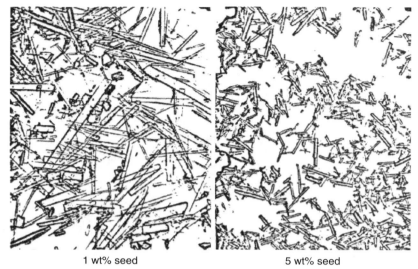

1 wt% seed   5 wt% seed

Figure 7-20   Final particle size using 1 wt% and 5 wt% seed.

Figure 7-21 shows the microscopic photos of the final product with and without sonication. As these figures shown, sonification breaks the crystals and drastically changes the crystal morphology, i.e., the "rod"-type crystals are changed to "cube"-type crystals.

EXAMPLE 7-4   *Resolution of Ibuprofen Via Stereospecific Crystallization*

Goals: Resolve S-ibuprofen-S-lysine from a pair of diastereomeric salts of R-ibuprofen-S-lysine and S-ibuprofen-S-lysine

Variables: L-S phase equilibrium, supersaturation, metastability, and crystal growth rate

Resolution via crystallization is an important method for the synthesis of pure optical isomers on an industrial scale. In the development of a process for the resolution of racemic ibuprofen, S-lysine is used as the resolving agent to accomplish this task. As shown in Fig. 7-22, R/S-ibuprofen reacts with S-lysine to form a pair of diastereomeric salts of R-ibuprofen-S-lysine and S-ibuprofen-S-lysine, and S-ibuprofen-S-lysine is separated via stereospecific crystallization.

Table 7-1   Experimental Conditions and Results of Crystallization

Experiment no.	Seed level, %	Seed surface area, m^2/gm	Product surface area, m^2/gm
270	3	>0.5	0.37
272	3	0.37	0.19
274-1 (with sonification)	3	0.19	0.34
274-2 (with sonification)	3	0.33	0.33

Without sonication   With sonication

Figure 7-21   Final particle size with and without sonication.

RESOLUTION OF IBUPROFEN WITH S-LYSINE

The method presented here is a semibatch process from the point of view of the slurry concentration, but it is semicontinuous by virtue of constant supersaturation and temperature driving force. Initially, S-ibu-S-lys seed is charged to the crystallizer and R/S-ibu-S-lys slurry is fed to the dissolver. During the course of resolution, S-ibu-S-lys is grown selectively in the crystallizer and R-ibu-S-lys is left in the dissolver. The resolution is achieved by the proper use of seed and control of supersaturation to grow the correct diastereomer.

LIQUID–SOLID PHASE EQUILIBRIUM

Table 7-2 lists the solubility of R/S-ibu-S-lys mixture in an ethanolic solution. Figure 7-23 plots these data on the ternary phase diagram. The data indicate that in the solid phase the tie lines join at the 100%

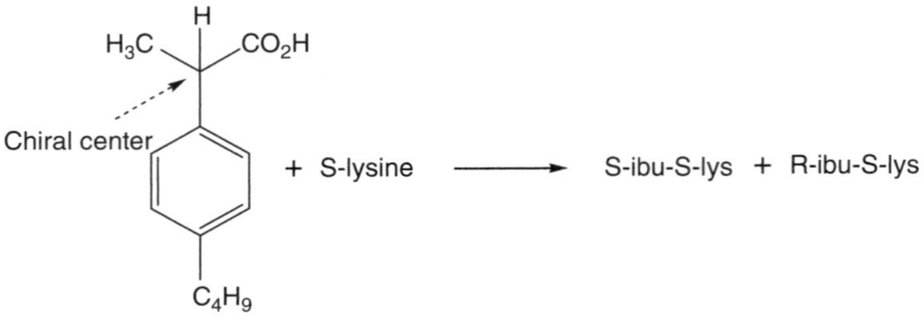

$$H_3C \underset{\text{Chiral center}}{\overset{H}{\diagup}} CO_2H$$

C_4H_9
R/S ibuprofen   + S-lysine $\longrightarrow$ S-ibu-S-lys + R-ibu-S-lys

Figure 7-22   Resolution of ibuprofen with S-lysine.

Table 7-2 Solubility of the R/S-Ibuprofen-S-Line in an Ethanol/Water 97/3 Solvent Mixture (23°C)

| Experiment | Solid phase | Liquid phase | |
	Purity, S-ibu %	Conc., mg/ml	Purity, S-ibu %
A	100	7.32	100
B	100	8.62	100
C	100	9.6	84
D	100	9.72	74
E	45	11.1	64
F	46	11	56.6
G	28	13.6	60

of S-ibu-s-lys (points A, B, C, and D). In the liquid phase, the tie lines join at an eutectic point of approximately 55% to 65% of S-ibu-S-lys (points E, F, and G). These data suggest that S-ibu-S-lys and R-ibu-S-lys crystals form a solid mixture rather than a compound. Therefore, resolution via crystallization is feasible.

METASTABLE REGION

A solution of R/S ibuprofen S-lysine diastereomers was found to sustain a high degree of supersaturation. The experimental observation showed that a supersaturated diastereomeric solution with a

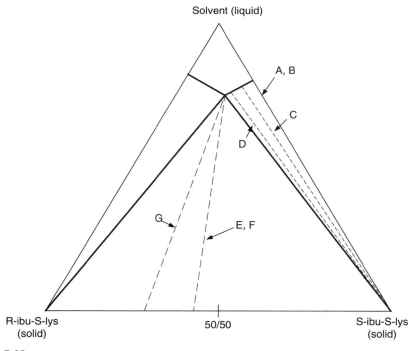

Figure 7-23 Solubility phase diagram.

supersaturation of ∼80 mg/ml remained clear after 16 hours without agitation. Additional tests indicated that seeding the supersaturated diastereomeric solution with S-ibu-S-lys did not prompt crystallization of R-ibu-S-lys after aging overnight.

CRYSTAL GROWTH RATE

During the batch crystallization which started with the same solution concentration of R-ibu-S-lys and S-ibu-S-lys (both are supersaturated, but R-ibu-S-lys is more supersaturated due to a lower solubility), the initial slurry sample shows a higher percentage of S-ibu-S-lys than R-ibu-S-lys. This indicates that S-ibu-S-lys has a higher crystal growth rate than R-ibu-S-lys. The difference in the crystal growth rate has a direct impact on the design of preferential crystallization. If the crystallizer is seeded with the diastereomer with a slower crystal growth rate, the seed bed is at greater risk of losing its optical purity if the wrong diastereomer starts to grow. On the other hand, if the crystallizer is seeded with one diastereomer that has a faster crystal growth rate, it will minimize the chance of optical contamination.

PREFERENTIAL CRYSTALLIZATION

To resolve the diastereomeric pair, the traditional approach is to apply the (equilibrium) solubility difference. This approach is not feasible in this case since the solubility of undesired R-ibu-S-lys is about two-thirds of that of desired S-ibu-S-lys. Thus, more R-ibu-S-lys than S-ibu-S-lys will crystallize.

Another option is to apply kinetic preferential crystallization, in which both isomers are supersaturated and the desired isomers are grown preferentially from the supersaturated solution. This approach overcomes the limitation of unfavorable equilibrium solubility. In order to apply kinetic preferential crystallization, one important issue to be considered is how to balance the unequal crystal growth rate of diastereomers. In preferential crystallization, balancing crystal growth rates between two isomers to be resolved is a prerequisite, since the starting material always contains an equal amount of two isomers. To maintain the overall steady-state material balance, both isomers must be grown and harvested at the same rate. For enantiomers, this requirement is easy to fulfill because enantiomers have identical crystal growth rates. For diastereomers, this requirement is difficult to meet because diastereomers have different crystal growth rates. Optical contamination can become an issue. When the diastereomer with the faster growth rate starts to nucleate/grow in the crystallizer that contains the other diastereomer, the optical purity of the crystal in the crystallizer will drop quickly. The design of a process to handle this unique property offers a true challenge.

Figure 7-24 presents a unique resolution configuration. This configuration provides intrinsically balanced crystal growth rates, avoids optical contamination, and resolves the diastereomers in a simple, direct manner. The essential feature is to seed the crystallizer with the faster-growing diastereomer and leave the slower-growing diastereomer in the dissolver. The crystal growth rates are balanced automatically by the overall material balance. Optical contamination is not an issue either.

Operationally, there are several ways to start up the system: (1) ibuprofen and lysine can be mixed in a separate tank/transfer vessel and then added or (2) the contents of the system at the end of one run can be saved for the next. It was decided to charge a slurry of diastereomers to be separated to the dissolver at the beginning. The slurry in the dissolver was continuously filtered via a ceramic cross-flow filter of 0.2 μm pore size. The supersaturated permeate was transferred to the crystallizer. Simultaneously, the slurry in the crystallizer could be filtered via another ceramic filter, and the clear saturated (with respect to S-ibu-S-lys) permeate filtrate could be sent to the dissolver. Both permeates would be kept at the same rate to maintain the volumes in both the dissolver and crystallizer. However, fluidized bed operation was clearly more convenient. Table 7-3 summarizes the results of two kilogram-scale experiments. As shown in the table, the final optical purity of S-ibu-S-lys is greater than 98%, starting with 50% S-ibu-S-lys and 50% R-ibu-S-lys in the dissolver.

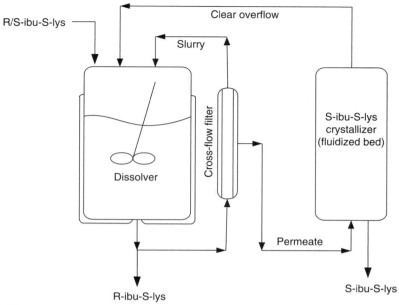

Figure 7-24 Flowsheet of preferential crystallization.

Table 7-3 Results of Resolution of Ibuprofen Lysinate

Run no.		292	296
Configuration		One dissolver (4 liter resin kettle with 0.2 μm 0.135 m² cross-flow filter)	
		One crystallizer (2 liter fluidized bed)	
Solvent		Ethanol/water 97/3, 6 liters	
Temperature		30°C for dissolver 25°C for crystallizer	
Flow rate		80–120 ml/min	
Operation time		70 hours	54 hours
Feed		~550 gm R/S-ibu-S-lys	~280 gm R/S-ibu-S-lys
Seed		50 gm S-ibu-S-lys, 99.4% optical purity in crystallizer	
Product (incl. seed)	Crystallizer	326 gm S-ibu-S-lys, 98.5% optical purity	194 gm S-ibu-S-lys, 98.1–98.8% optical purity
	Dissolver	269 gm R-ibu-S-lys, 92–100% optical purity	130 gm R-ibu-S-lys, 93.4–98% optical purity
Crystallizer yield (corrected for seed)		276 gm (100%)	144 gm (100%)

> **EXAMPLE 7-5**   *Crystallization of Pure Bulk with Polymorphism*

Goals: Formation of the correct crystal form

Variables: Polymorph solubilities, control of supersaturation, crystal growth rate

As discussed in Chapter 1, polymorphism is a common problem encountered in the synthesis of pharmaceuticals. In the process development for a drug candidate of reverse transcriptase inhibitor, six crystal forms were identified. The first pilot plant batch produced all Form III, not the desired Form I. The purpose of this example is to illustrate the development of a robust crystallization process to consistently grow the desired crystal form.

BRIEF PROCESS DESCRIPTION

The process illustrated here is a simple modification of a previous example. Shown in Fig. 7-25, this modified process prepares the crude drug as a slurry in a feed vessel. The slurry feed was continuously charged to a dissolver that was maintained at a temperature of about 50°C. The feed rate was controlled such that the slurry was put into solution in the dissolver, and the dissolved solution charged continuously to the crystallizer through an in-line filter to remove extraneous insoluble particles and traces of the undissolved product. The crystallizer contained the seed slurry with the correct form at a lower temperature, about 25°C. The crystallizer slurry was continuously filtered through a ceramic cross-flow filter system with a pore size of 0.2 μm, and the clear permeate was sent back to the dissolver for further solubilization of product. This was run until the feed tank was empty and all supersaturation was relieved. The critical parameters for successful development of this process were solubilities of the polymorphs, seeding, and control of supersaturation.

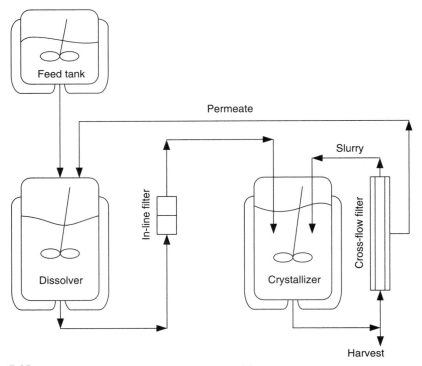

Figure 7-25   Flowsheet of crystallization with control crystal form.

SOLUBILITIES OF POLYMORPHS

As mentioned earlier, a total of six crystal forms were identified with this drug candidate. The solubilities of Forms I and III are shown in Fig. 2.10. It is likely that Form I is the most stable form at room temperature, whereas Form III appears to be the most stable form at a temperature above 75°C. Based upon the data, it is clear that the solution concentration during the crystallization should not exceed the solubilities of undesired forms in order to prevent the formation of undesired crystal forms. Additionally, there is an upper limit on the operating temperature and the solution concentration due to the presence of enantiotrophic Form III.

SEEDING AND CONTROL OF SUPERSATURATION

To provide the best practical environment for growing the correct crystal form, the semicontinuous crystallization process shown in Fig. 7-25 was successfully developed and demonstrated in the pilot plant. Table 7-4 summarizes the operating conditions and the results of laboratory and pilot plant operation.

In order to grow the correct crystal form, seed crystals of Form I were initially charged to the crystallizer. Dissolver and crystallizer temperatures were maintained at 50°C and 25°C, respectively, so that the concentration in the crystallizer was below the solubility of the other forms. Production rate was 5 gm per hour per liter of crystallizer. Under these conditions, the slurry concentration in the crystallizer started at 10 mg/ml and ended at ~100 mg/ml after 20 hours of operation. As shown in Table 7-4, this process successfully produced the correct Form I without the contamination of other undesired forms.

EXAMPLE 7-6   *Continuous Separation of Stereoisomers*

Goals: Kinetic separation of isomers by crystal growth of the desired isomer on a seed bed of that isomer (desired: all-growth process)
Robust continuous process applicable to selective crystallization in general

Issues: Minimize nucleation of undesired isomer—strict control of supersaturation.
Maintain particle balance—create new growth centers in the absence of nucleation.

This process is outlined here as an example of the use of tightly controlled supersaturation in a cooling crystallization. It is discussed in greater detail in Example 11-6.

Many pharmaceutical products are specific stereoisomers, and in many cases the inactive enantiomer must be removed to minimize side effects. In recent years, advances in selective stereochemistry

---

Table 7-4   Experimental Conditions and Results of Polymorphs

| Run no. | Crude, kgs | Seed, kgs | Yield, % | Product purity, wt% | Crystal form | Particle size distribution, % less than 25 μm | |
						Seed	Batch
Lab. 37A	2.21	0.3	91	99.6	I	93	89
Pilot Plant 3-1	22.2	2.2	78	99.5	I	90	85
Pilot Plant 4-1	25.5	3.2	86	100	I	85	74

have reduced the need for isomer separation. However, many synthetic processes continue to have racemic output (desired and undesired stereoisomers).

Separation of these products reduces the possibility of side effects from the inactive isomer. In addition to the reduction of side effects, separation of these isomers often makes possible a greatly improved yield if the inactive entity can be racemized and recycled.

Because enantiomers of the same compound have many of the same thermodynamic properties, including solubility characteristics in nonchiral solvents, they are particularly difficult to separate in equilibrium processes.

SOLUTION

- Kinetic separation of enantiomers in heavily seeded, all-growth, limited residence time crystallizers.
- Continuous (steady-state) process for tight control of supersaturation, unchanging with time.
- Fluidized bed crystallizers to avoid the need for heavy-magma, high-flux filtration equipment.

A brief discussion of processes introduced by Merck & Company for "kinetic" resolution of stereoisomers by crystallization is included here. A more detailed description can be found in Chapter 11.

FLOW PATTERNS

Typical flow diagrams for both stirred tank and fluidized bed resolution systems are shown in Figs. 7-26 and 7-27. They are shown, respectively, as series flow for the CSTR crystallizer system and parallel flow for the fluidized bed operation, but the decision to run in series or parallel is affected only slightly by the crystallizer type.

The advantage of series operation is that the upstream crystallizer can remove some of the supersaturation of the undesired isomer prior to entering the downstream crystallizer, which is always the one containing the desired final product.

Parallel operation does not protect the desired product from optical contamination, but it is easier to run because both crystallizers are independent of each other. Neither flow reduction in one unit from product withdrawal nor inadvertent washout of seed crystals has a significant impact on the other and generally causes parallel flow to be the system of choice.

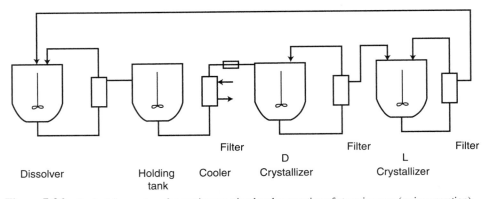

Figure 7-26   Typical flow pattern for continuous stirred tank separation of stereoisomers (series operation).

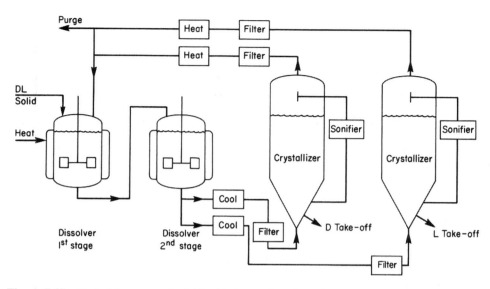

Figure 7-27 Typical flow pattern for fluidized bed separation of stereoisomers (parallel operation).

CRYSTAL GROWTH

Growth rate kinetics for most of the stereoisomer resolution processes run at Merck have been essentially first-order with respect to supersaturation, although there have been exceptions (see Chapter 11). For one of the stereoisomer separation systems, the (first-order) growth rate constant was measured for both the stirred tank (CSTR) and fluidized bed flow systems, with and without a nucleation suppressant. The data are shown in Table 7-5.

In this case at least, the fluidized bed system showed considerably faster growth rate kinetics than the stirred tank system, presumably by convective enhancement in the boundary layer on the surface of the crystals. Additionally, since nucleation suppressants nearly always also inhibit the growth rate, the slower growth kinetics in this case are not unexpected.

POPULATION BALANCE CONTROL

The astute observer will see that the fluidized bed crystallization system shown in Fig. 7-27 has an unusual feature. Flow sonication units are in position to operate on pumped slurry from the seed beds. In other systems, the sonicators are located internally in the bottom of the column. The sonicators are

Table 7-5 First-Order Growth Rate Constants (Fluidized Bed versus CSTR Crystallizer System) $k \times 1000$: gm/hr, cm^2, gm/cm^3

Crystallizer temperature, °C	Fluidized bed		CSTR
	No nucleation suppressant	Nucleation suppressant added	Nucleation suppressant added
25	40		2.2
35	70	16	2.3
41	90		2.3
46			4.1

power ultrasonic horns, which create sufficient cavitation and impact energy to break the crystals along cleavage planes.

Sonication is the means used in the fluidized bed crystallizers to maintain the number of seed particles in the magma to replace those removed in the product and at the same time to prevent the formation of overly large crystals. Excessive particle size starves the seed bed of crystal surface area for growth and, in the case of fluidized bed crystallizers, causes sluggish solids movement, which can cause the particles to grow together.

Figure 7-28 shows the effect of sonication on crystal seed particles in one fluidized bed operation. The unsonicated crystals (top photo) are disrupted along cleavage planes (bottom photo), a mechanism which produces few fines to elutriate out of the seed bed. The sonicators are located internally or in a recycle system, but the crystal population fed to them is always drawn from the bottom of the seed bed where the largest particles are located.

Figure 7-29 shows another product with more needle-shaped morphology, which is externally sonicated in a recycle flow loop. Seed crystals are shown entering (top photo) and leaving (bottom photo) the flow sonication unit.

FLUIDIZED BED CRYSTALLIZER SCALE-UP

When scaling up column geometry, it is desirable to know the localized slurry concentrations within the crystallizer. Excessively high concentrations can result in sluggish fluidization and possible agglomeration, and excessively low concentrations decrease productivity. These local concentrations

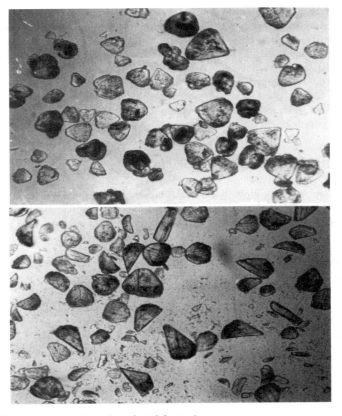

Figure 7-28   Unsonicated crystals (top); sonicated (bottom).

Figure 7-29 Crystals entering (top) and leaving (bottom) a flow sonication unit.

can be calculated, assuming a classified bed (largest particles on the bottom) and either a measured or estimated size distribution. The results from such a calculation are shown in Chapter 11.

To estimate the amount of ultrasonic energy required to run in the scaled-up crystallizer, a study was carried out measuring crystal disruption by elutriation from a glass flow unit. As shown in Table 7-6, breakage was essentially first-order with sonication power and slurry concentration. This also implies that the main mechanism for crystal breakage is individual crystal breakage in the

Table 7-6 First-Order Rate Constants for Ultrasonic Crystal Breakage $\dfrac{1}{P}\dfrac{dm}{dt} = k'$ (slurry − concentration)

P (watts)	k' [gm/hr per (gm/liter, watt)]
30	0.031
55	0.028
68	0.029

Figure 7-30   Factory fluidized bed crystallizers.

cavitation field rather than particle-particle interaction, which is more common for most wet milling operations. The "breakage constant" of about 0.03 gm/hr per (gm/liter, watt) was found to hold for a number of small-molecule organic compounds tested in the fluidized bed crystallizer system.

Figure 7-30 shows a pair of factory fluidized bed crystallizers, one for each stereoisomer, in construction. Internal sonicators were installed in the column bottoms. The blowers shown at the bottom of each column were used to cool the sonication units. Calculated values of localized solids concentration and PSD in these columns were in good agreement with predicted values from the calculation procedure described above.

Chapter 8

Evaporative Crystallization

8.1 INTRODUCTION

Increasing the concentration by evaporation or distillation is a common method of increasing supersaturation and inducing crystallization. Since solvent is removed over a finite period of time, it is inherently a semibatch operation. Semicontinuous or continuous operation is also possible. The evaporation or distillation can be run at atmospheric pressure, or at reduced pressure when substrate stability is not compatible with the required atmospheric distillation temperature.

One of the primary advantages of evaporative procedures is that they can often be combined with other process operations to reduce equipment requirements and/or time cycles. In addition, it is possible in some cases to complete the crystallization without the addition of a second solvent, thereby avoiding the costs of separation and recovery. Some of the process advantages that may be realized are as follows:

- combination with a change in solvent and simultaneous crystallization
- combination with reaction and removal of a volatile reaction by-product and simultaneous crystallization
- combination with cooling crystallization
- combination with antisolvent crystallization

These operational advantages must be evaluated against the disadvantages that are discussed in the sections to follow. These disadvantages may include

- difficulty in controlling mean particle size and particle size distribution (PSD)
- difficulty in determining the seed point
- unpredictability on scale-up
- inconsistent batch-to-batch performance

8.2 SOLUBILITY DIAGRAMS

Figure 8-1 shows the path of concentration versus time in relation to saturation solubility. With the initial concentration at point A, solvent is removed until the concentration crosses the equilibrium solubility line at point B, where crystallization may occur. In most cases, however, some degree of supersaturation is necessary before crystallization actually starts,

Crystallization of Organic Compounds: An Industrial Perspective. By H.-H. Tung, E. L. Paul, M. Midler, and J. A. McCauley

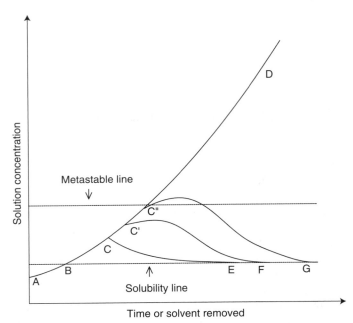

Figure 8-1   Concentration profiles for crystallization by evaporation as a function of time of distillation or amount of solvent removed. A-B-C-E is the preferred pathway for favoring growth.

depending on the width of the metastable region, the presence or absence of seed, and other factors such as mixing, bubble formation, and impurities level. Once crystallization is initiated, the concentration in solution may follow different pathways. Assuming that crystallization begins at point C, the pathway between points C and E depends on (1) the rate of distillation, (2) the surface area of the crystals, (3) the secondary nucleation rate, and (4) the inherent crystal growth rate. In the absence of crystallization, the concentration would reach point D, the final overall concentration.

Figure 8-1 shows additional concentration pathways (B-C'-F, B-C''-G). Crystallization can begin at different concentrations C' and C'', depending on the factors mentioned above. Once crystallization is initiated, the degree of supersaturation, as indicated by the departure of the actual concentration profile from the saturation curve (B-E-F-G), that is actually achieved between initial crystallization at C' or C'' and termination (E, F or G) is determined by the seed area, inherent crystal growth rate, and secondary nucleation. In addition, oiling out, agglomeration, and/or extensive nucleation can occur if the concentration profile goes beyond the metastable region (above C''). These issues are discussed in Chapter 5 and can occur in any crystallization operation at supersaturations above the metastable region.

8.2.1   Increasing Solubility

Two factors complicate these simple representations. They are the effect of increasing distillation temperature caused by the increasing concentration of substrate and impurities. The impurities, in particular, can dramatically increase the solubility of the substrate. They can also decrease the inherent growth rate by blocking or inhibiting surface incorporation on the growing crystals or by reducing the nucleation rate. These effects are represented as curve H-E in Fig. 8-2. As distillation proceeds, the solubility increases. In extreme cases, the crystals once formed could melt as the temperature increases.

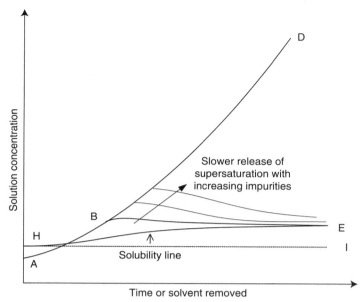

Figure 8-2 Concentration profiles for crystallization by evaporation when the solubility of the substrate is increasing because of increasing temperature and/or impurity concentration. The solubility profile is no longer curve H-I but becomes H-E.

Increasing solubility because of increased concentration of impurities will result in a similar equilibrium change, although in some cases, the effect could be much greater. In extreme cases, when the residual solvent concentration is reduced to less than a critical value, the substrate could melt or solidify, depending on the melting point and the impurity effect. This condition is often used in laboratory preparations for convenience in changing solvents and is referred to as concentration to dryness. It is obviously not a scalable operation in a stirred vessel. Specialized tubular evaporators with close-clearance or scraped-surface rotors are available for these applications and have been successfully used by the authors for concentration but not for simultaneous crystallization.

The actual concentration curves that could result are shown in curve B-E. Obviously, a yield loss will result because of the increased equilibrium solubility.

8.2.2 Decreasing Solubility

A less common effect of impurities is a decrease in the solubility of the substrate. This effect could be attributed to an impurity that is in high concentration because it will not crystallize under the existing conditions or because the temperature is above its melting point. The solvent capacity for the substrate is thereby decreased, resulting in decreased solubility. In aqueous systems, this effect could be caused by high inorganic salt concentrations and may be referred to as salting out.

8.2.3 Change in Solvent

A common variation of the evaporative operation is to evaporate the original solvent and charge a second solvent with reduced solubility for the substrate. This can be accomplished either in batch or semicontinuous mode. The solubility will then decrease as distillation

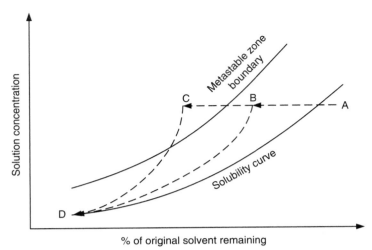

Figure 8-3   Concentration profiles for crystallization by evaporation when the batch volume is held constant while an antisolvent is being added.

proceeds and the system becomes richer in the second solvent. Supersaturation is generated due to the change in equilibrium solubility at different solvent compositions. The temperature will also increase or decrease accordingly, depending upon the boiling points of the solvents.

All of the above control factors are issues, as well as the local concentration gradients at the point of addition of the second solvent (also referred to as the antisolvent). This key point will be discussed in Chapter 9. Also discussed in that chapter is the method of seed introduction as a slurry with the antisolvent. This method can be useful in this case as well.

A change in solvent can be operated at essentially constant volume by balancing the distillation rate with the feed rate of the second solvent being added. In this case, the solution concentration profile will resemble closely that of cooling crystallization, as shown in Fig. 8-3. The amount of the original solvent remaining is shown as the abscissa. The evaporation is originated at point A and the antisolvent addition is started at the same time at a rate to maintain constant volume. By choosing the correct rate of evaporation and by seeding at point B, the concentration can adhere closely to the equilibrium saturation and end at point D. If the evaporation rate is too fast or seeding is delayed, the concentration can exceed the metastable limit (point C) and the concentration will then follow (C → D). Examples are given below to highlight the impact of distillation on supersaturation and PSD.

8.3   FACTORS AFFECTING NUCLEATION AND GROWTH

One of the primary difficulties with distillation as a method of inducing crystallization is control of nucleation. The factors affecting nucleation are the potentially large local gradients in both concentration and temperature that are induced by the generation of vapor at the heating surfaces and at the liquid-vapor interface as the vapor bubbles leave the liquid surface. Local concentration can be greater than the limits of the metastable region, thereby causing increased local nucleation rates. The result on the ultimate mean particle size and PSD can be much less predictable. Both of these results can be affected by the distillation rate, wall temperature, vapor disengagement, and vapor bubble distribution as well as by the factors discussed above.

Distillation is analogous to the cooling rate in creating supersaturation and may be controlled by similar methods to match the evaporation rate with the surface area available for growth. However, the wall and vapor disengagement effects on local concentration can cause excessive nucleation such that a predictable growth rate may not be achieved. Other factors that are difficult to control are as follows:

- build-up of product scale at the evaporation surface and above the surface on the heated wall (see Section 8.5.1 below)
- decomposition of product in the scale due to local overheating
- occlusion of impurities and/or solvent in the crystals due to a local rapid nucleation/growth rate
- determination of the seeding point during evaporation
- poor bulk mixing caused by increasing slurry concentrations
- foaming (can be increased by the presence of fine particles)

8.4 SCALE-UP

All of the noted problems with control of nucleation and growth are exacerbated on scale-up, thereby increasing the risk of an unpredictable PSD and other physical attributes. The causes of these scale-up problems include

- reduced surface area to volume ratio for heating
- reduced relative distillation rate and/or higher jacket/coil temperature
- reduced vapor disengagement area relative to volume—increased foaming
- increased mixing turnover time
- change in secondary nucleation due to changes in mixing speed, shear, micro- and macro-mixing

These factors may have conflicting effects on PSD. They are difficult to quantify and are system specific, depending primarily on the crystallization characteristics of the substrate. Therefore, laboratory results can vary greatly from pilot scale results and from pilot scale to manufacturing scale.

The net result of these control issues is that crystallization by evaporation of solvent may be satisfactory for intermediates when control of PSD and other physical factors may not be a significant issue (with the possible exception of downstream operations such as filtration). However, for final bulk drug products where control of these attributes is critical, this method has proven to be unsatisfactory.

8.5 EQUIPMENT

Evaporative crystallization can be carried out in a wide variety of equipment for both semi-batch and continuous operation. Discussions of these systems may be found in several references including Myerson (2002, chapter 10) and Mullin (2001, chapter 7). They are utilized in some areas of the chemical industry that require the high production rates that can be achieved with these specialized designs.

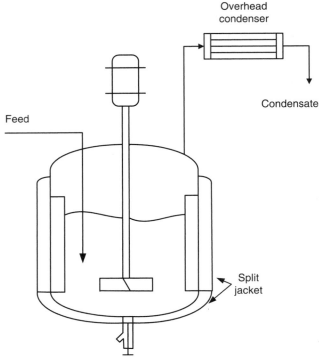

Figure 8-4 A standard jacketed stirred tank with baffles and an overhead condenser for evaporative crystallization. During distillation, a second stream of solvent can be charged to maintain a constant batch volume in the vessel.

Our experience in the pharmaceutical industry has been confined to semibatch operation in standard stirred vessels, as illustrated in Fig. 8-4. These vessels may be operated at atmospheric pressure or more commonly under reduced pressure, for which vacuum pumps are the most common vacuum source.

As indicated in the discussions above, these applications may be subject to difficult control issues on scale-up. Impeller type and power input can be difficult to specify because of the conflicting requirements of providing adequate circulation without breaking crystals with excessive shear and/or creating excess nucleation, as discussed in Chapter 6.

8.5.1 Heat Transfer

In addition to these standard crystallization issues, evaporative crystallizers must have adequate circulation at the wall surface and maintain sufficient temperature difference between the jacket fluid and batch fluid for heat transfer to achieve satisfactory distillation rates. This issue can be complicated by restrictions on wall temperature imposed by the common temperature instability of organic compounds. These problems can be severe when the crystallizing material forms a crust on the vessel wall, where it can decompose as well as reduce heat flux.

The possible restriction on jacket temperature can preclude the use of a common jacket service such as pressure steam. An alternative is hot water or a heat transfer fluid at a temperature compatible with the decomposition limits. Both of these liquids result in decreased jacket side heat transfer coefficients compared to those for steam. Another possible alternative is steam under vacuum, which can be readily achieved with a vacuum pump and can be used to limit jacket temperatures to ~60°C while maintaining good wall heat transfer rates.

8.5.1.1 Split Jacket Vessels

Perhaps the most important and most readily achievable equipment design to minimize wall decomposition is the use of jacket services on the bottom section of the vessel only. Such services can be provided either by a lower jacket only or, more commonly, by a split jacket to provide for full services in multipurpose vessels. Decomposition can be especially severe on the wall above the boiling surface, where any liquid may evaporate and leave the residual compound to dry and be exposed to the maximum temperature in the jacket (baking).

The best defense against these possibly severe problems of product loss and contamination with impurities is a design and an operation that preclude heat supply near or above the boiling surface at all times. This precaution may be especially important as the batch volume decreases and baking on the wall may be severe.

Split jacket designs are available for virtually all vessels, and the additional cost may be offset by the savings realized in preventing decomposition at the upper wall. In addition, by limiting exposure at the critical liquid-vapor interface, the temperature of the heating fluid

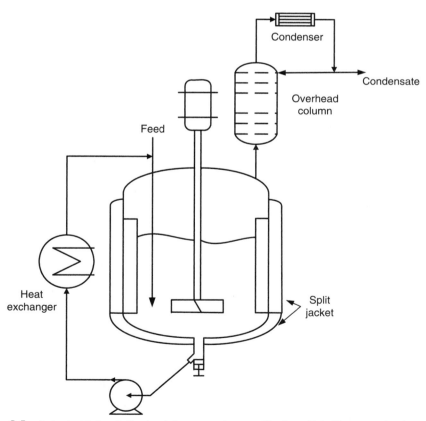

Figure 8-5   A standard jacketed stirred tank for evaporative crystallization with baffles, an overhead condenser, and an external heat exchanger to improve heat transfer. Similarly, during distillation, a second stream of solvent can be charged to maintain a constant batch volume in the vessel. Fractionation is a viable option for aiding solvent change and/or recovery.

may be increased at the submerged heating surface and the additional driving force may compensate for the decrease in overall heat transfer surface area.

8.5.1.2 *External Heat Exchangers and Internal Coils*

In some cases, the required distillation rate can be achieved by forced circulation through a heat exchanger, as shown in Fig. 8-5. This design can achieve higher heat input at a minimum temperature difference by providing more surface area than the vessel jacket and higher heat transfer coefficients. However, the crystals are subject to additional shear stresses in the pump and possible encrustation on the heat exchanger tube surfaces, thereby limiting this design to robust systems.

Internal coils can also be used to increase heat transfer by supplying additional surface area. However, the coil surfaces may be subject to encrustation and may be difficult to clean.

8.5.1.3 *Alternatives to Evaporation*

In cases of extreme temperature instability, evaporative methods of concentration have not been feasible because distillation temperatures at even the lowest feasible pressures are still too high (e.g., distilling water at 25°C may be possible but this temperature may be still too high for the extended time cycle required). Alternative means of concentration are reverse osmosis for aqueous systems and some membrane systems with molecular weight cutoffs compatible with the compound in question. Obviously, these systems preclude crystallization during operation. An example of an antibiotic with severe temperature restrictions is presented in Example 11-1.

8.5.2 Overconcentration

The potentially most damaging operational error is overconcentration. As the solids concentration in solution and/or in the crystal slurry increases, the viscosity and temperature can increase. This condition can most likely occur as the residual volume reaches the region of the impeller and perhaps—in extreme cases—below the impeller. Possible consequences when a critical concentration is reached include

- solidification of the entire mass
- decomposition of the substrate
- damage to the mixing components

In extreme cases, the temperature could increase to the temperature of the heating fluid and initiate a decomposition exotherm in the residual mass, causing a catastrophic failure of the vessel.

Measures to prevent such events include

- determination of the decomposition initiation temperature
- limiting heating fluid temperatures to a safe margin below this temperature
- provision of control measures to detect the approach to a critical temperature, concentration, and/or volume limit
- provision of shutdown procedures if this temperature, concentration, and/or volume are reached (by cooling but not breaking the vacuum)

8.5.3   Combination of Evaporation and Cooling

In addition to the common combination of evaporation and antisolvent addition, evaporation is often combined with cooling to provide maximum yield. In some cases, the substrate does not crystallize during concentration, either because the solubility at the higher distillation temperature is greater than the solution concentration or because of a wide metastable zone allowing high supersaturation. In these cases, the issues discussed in Chapter 7 on cooling become important when the concentration is complete and cooling is initiated. Cooling a solution that is highly supersaturated can readily result in oiling out, agglomeration, and/or crash-out.

In most cases, crystallization occurs during concentration and the precautions outlined in this chapter are relevant.

EXAMPLE 8-1   *Crystallization of a Pharmaceutical Intermediate Salt*

Goals: To grow large crystals with acceptable filtration properties

Issues: Controlled seeding and distillation rates

In the production of an intermediate compound in the synthesis of a pharmaceutical product, filtration difficulty was encountered after the crystallization of that intermediate by distillation. The filtration problem was attributed to the presence of very fine needles (~2 microns thick).

OPTION 1

In the initial procedure, after reaction and extraction, the intermediate was present in an organic layer containing THF, toluene, hexane, and some residual water. The mixture was evaporated until the tetrahydrofuran (THF) was essentially completely removed, as shown in Fig. 8-6 (A-B). Seed was added at this point, and the crystallization was allowed to go to completion (B-C). (The width of the

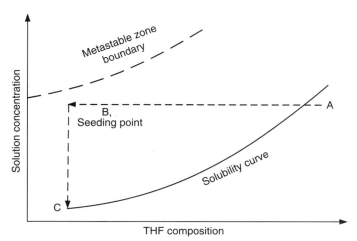

Figure 8-6   A concentration profile for uncontrolled crystallization by evaporation (initial procedure in Example 8-1. The seed was charged to a highly supersaturated solution which may exceed the metastable zone boundary. The metastable zone width was not measured, so this figure is used only for illustration purposes.

metastable zone was not measured, so the representation shown in Fig. 8-6 is for illustration purposes only.) The resulting filtration was unacceptably slow because of the 2 micron crystals that resulted from the high supersaturation at the time of seeding.

OPTION 2

In the modified procedure, the organic solution was distilled under vacuum until the batch concentration exceeded the solubility limit. In order to avoid undesired spontaneous nucleation, seed was charged at this point and the batch was aged to develop a seed bed. Following the seed age, toluene was fed and THF was distilled continuously under vacuum while maintaining the same batch volume. On completion of the distillation, additional hexane was added to minimize product loss in the mother liquor. The key parameters for this process are construction of the solubility map, determination of the seeding point, and control of the solvent evaporation rate to release the supersaturation at the proper rate.

The impact of solvent composition on solubility is shown in Fig. 8-7. As shown in this figure, solubility is proportional to the THF level in the solvent mixture. Along the solubility curve, the data also highlight the corresponding hexane percentage because the system is a quaternary solvent system containing THF, toluene, hexane, and some residual water. Upon distillation, the hexane level is reduced simultaneously with the THF.

Figure 8-7 further highlights several key action steps during crystallization: initial condition A, seeding point B, end of aging point C, and end of distillation point D. The composition moves from point A to point B during the first distillation The solution concentration was kept the same under constant volume distillation. At point B, the batch was seeded and a seed age time added to release the supersaturation (B-C). Once it reached point C (verified by sampling), the composition was changed as shown from point C to point D during a second distillation. The distillation rate was controlled so that the solution concentration profile closely follows the solubility profile, which was reduced gradually. A total of 10 hours was taken for the second distillation.

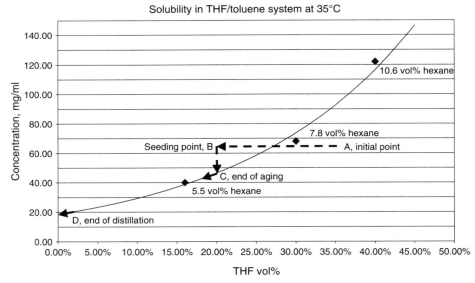

Figure 8-7 A concentration profile for a controlled crystallization by evaporation (revised procedure) with seeding and a seed age in Example 8-1. The seeding point was much closer to the solubility curve, and sufficient seed aging was given to release the supersaturation before further distillation.

In this option, the batch is seeded at much lower supersaturation to avoid excessive nucleation, as experienced in Option 1 at higher supersaturation. The modified approach was successfully scaled up from the laboratory 50 gm scale to pilot plant and manufacturing 200 kg over 4000× scale-up factor. The crystals size increased to an average length of ~60 microns and a width of ~4 microns, compared to the Option 1 procedure that produced crystals with a length of ~25 microns and a width of ~2 microns. The filtration rate was improved substantially from several days' operation to less than 8 hours in the pilot plant.

EXAMPLE 8-2   *Crystallization of the Sodium Salt of a Drug Candidate*

Goals: To demonstrate the impact of controlled crystal growth rate on crystal morphology

Issues: Crystal morphology, continuous crystallization

In the crystallization of the sodium salt of a drug candidate, extremely fine needles were generated. In addition, changes in crystal morphology and crystallinity were noted after filtration and drying. In order to avoid these and to grow better crystals, an alternate continuous evaporative crystallization approach with heavy seeding was evaluated.

OPTION 1

In the batch-mode operation, the conversion of free acid to the sodium salt was carried out in an ethanol/acetone mixture. The sodium salt solution was prepared in the ethanol solution. The antisolvent acetone was added to the batch at 45°C to 50°C, with a trace amount of seed added to promote crystallization. The initial solids crystallized in the solution were a mixture of crystalline and amorphous materials. After extended aging, the amorphous material converted to needle-shaped crystalline material. The slurry was cooled, filtered, and washed with acetone. The wet cake was vacuum dried at 50°C.

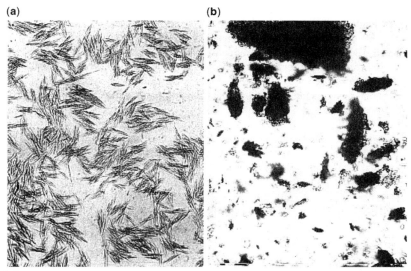

(a)   (b)

Figure 8-8   Crystals of batch-mode crystallization (a) as a slurry and (b) as the dry cake of Example 8-2. Note the bundles of fine needles.

(a)   **(b)**

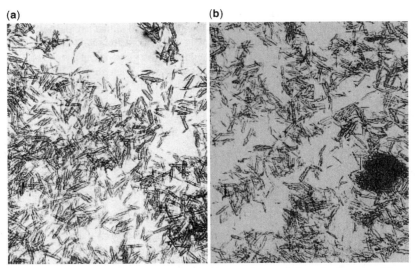

Figure 8-9   Crystals of semicontinuous crystallization (a) as a slurry and (b) as the dry cake of Example 8-2. Note that the needles are more three-dimensional and less bundled than those in Fig. 8-8.

In the pilot plant, intermittent delumping of the partially dried cake was needed to reach the desired dryness. A change in crystal morphology was noted during the filtration and drying stages.

As shown in Fig. 8-8, the appearance of crystals suggests that the crystal growth rate in this operation was fast. Furthermore, the dried cake appears to be lumpy and less crystalline.

OPTION 2

In the proposed continuous evaporative crystallization approach, a fixed amount of solid compound obtained from a batch crystallization was slurried in a crystallizer in a solvent mixture of ethanol and *n*-butyl acetate. (*n*-Butyl acetate replaced acetone as the antisolvent.) *n*-Butyl acetate has a higher boiling point than ethanol. Therefore, as described below, ethanol can be readily removed by evaporation.

In a separate feed vessel, a homogeneous solution with compound dissolved in ethanol was prepared. The homogeneous ethanolic solution was charged continuously to the crystallizer at a rate which was balanced by the vacuum distillation rate of ethanol in the crystallizer. The rate of batch addition and distillation was designed so that the overall crystal growth rate in the crystallizer was approximately 1% per hour, i.e., growing 1 gm of material from 100 gm of seed bed per hour. The crystallizer slurry was periodically harvested.

Crystals generated by the continuous evaporative crystallization are shown in Fig. 8-9. As can be seen, continuous evaporative crystals are much thicker. Furthermore, no change in crystal morphology after filtration and drying was observed. This clearly indicates that the thicker crystals are better than the original batch crystals as a result of a carefully controlled crystal growth environment in the presence of heavy seeding and a low growth rate.

A batch size up to 100 gm was prepared in the laboratory. Large-scale production, including piloting, was not initiated due to a change in the clinical program.

MESSAGE

Continuous crystallization with control of the crystal growth rate can significantly improve the crystal morphology.

Chapter 9

Antisolvent Crystallization

Addition of an antisolvent is potentially the best method to achieve controlled and scalable particle size distribution (PSD). The addition can be either semibatch—antisolvent added to product solution or product solution added to antisolvent (reverse addition)—or continuous using an in-line mixer or stirred vessel. Of these, the in-line mixer offers in some cases the highest potential for predictable results, as will be discussed below.

The obvious disadvantage of the antisolvent process is the necessity to introduce an additional solvent or solvents, thereby reducing the volumetric productivity and creating a solvent mixture requiring some form of purification/separation for downstream processing and/or recovery.

9.1 SEMIBATCH OPERATION

9.1.1 Normal Mode of Addition

The addition of an antisolvent can be carried out in different ways, as indicated in Fig. 9-1, where the concentration of product is shown on the ordinate and the amount of antisolvent added is shown on the abscissa. A typical equilibrium solubility curve is indicated as A-B-C. (This curve could be concave or linear but is shown as convex for clarity.) The metastable region is indicated as the area between B-C and E-D. From point A to point B, addition of antisolvent will proceed without crystallization because the solution concentration is below the equilibrium solubility. At point B, equilibrium solubility is reached. As the addition of antisolvent continues, supersaturation will develop. The amount of supersaturation that can be developed without nucleation is system specific and will depend on the addition rate, mixing, primary and/or secondary nucleation rate, and growth rate, as well as the amount and type of impurities present in solution.

If growth-dominated crystallization is desired, with the presence of a sufficient quantity of seed and a sufficiently slow addition rate, the concentration in solution may remain completely in the metastable region as crystallization proceeds. The closer the solution concentration profile is to the equilibrium solubility curve (B-C), the higher the possibility of achieving an all-growth process.

On the other hand, a system without seed or a high addition rate can develop a high degree of supersaturation, which can result in rapid precipitation or crash-out at point B'', beyond the metastable zone. Crash-out could be followed by continued nucleation and some growth (B''-C), and eventually to equilibrium at some time after all the antisolvent

Crystallization of Organic Compounds: An Industrial Perspective. By H.-H. Tung, E. L. Paul, M. Midler, and J. A. McCauley
Copyright © 2009 John Wiley & Sons, Inc.

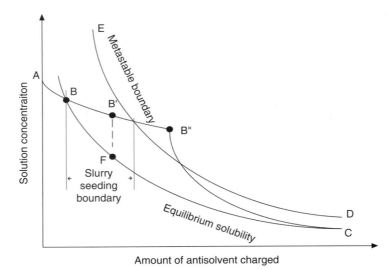

Figure 9-1   Concentration profiles for normal addition of antisolvent to batch solution with reference to equilibrium saturation and the metastable zone. This figure shows various points along the path of addition of antisolvent. If the rate of addition of antisolvent is too fast and without seeding, the solution concentration may exceed the upper boundary of the metastable zone and solute may crash out. On the other hand, if the solution is seeded and aged within the metastable zone, growth may be enhanced, with reduced nucleation.

is added. When the concentration is allowed to go to point B″, the system is also subject to oiling out and/or agglomeration, as discussed below in Section 9.1.2.

A common practical approach to achieving growth while minimizing concern for seed dissolution is shown in curve B′-F-C. The addition of antisolvent is stopped, and seed is added at point B′, where the system is slightly supersaturated. As discussed in Chapter 2, in-line measurement by Fourier transform infrared (FTIR) or ultraviolet (UV) can be utilized to determine when the concentration reaches point B′. Alternatively, the seed may be added in a slurry with the antisolvent starting before point B is reached, as discussed in Chapter 5 and below in Section 9.1.4.

Crystallization is then allowed to proceed to relieve the supersaturation without the addition of more antisolvent (B′-F). Given sufficient time, point F could closely approach the equilibrium solubility value while developing sufficient surface area to achieve essentially all growth when addition of antisolvent is resumed. Given this increased surface area and a sufficiently slow addition rate, the solution concentration can approach the equilibrium solubility for the remainder of the addition (F-C).

9.1.2   Reverse Addition

In the event that fine particles are desired, the order of addition can be reversed and the product solution added to the antisolvent. This method is shown in Fig. 9-2. The solubility in the antisolvent is very low at point A. Because of the low solubility, the supersaturation ratio will increase rapidly (E-F) before there is sufficient seed area to achieve any significant growth. If the addition is fast enough, growth can be eliminated, resulting in all nucleation and creation of very fine particles. Even with slower addition and seed initially present, the low

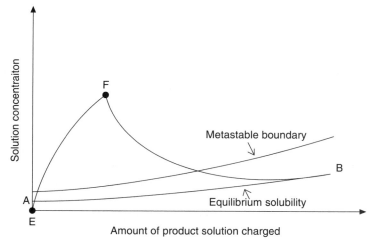

Figure 9-2 Concentration profile for reverse addition (solute solution to antisolvent) with reference to equilibrium saturation and the metastable zone. As solution is added to the antisolvent, the solute concentration can rapidly exceed the solubility limit and the metastable zone boundary, resulting in significant nucleation.

equilibrium solubility can create high supersaturation ratios throughout the addition and result in the generation of small particles.

A potential complication for all crystallization methods, and particularly for this type of addition, is the tendency for organic compounds to oil out and/or agglomerate as fine particles into amorphous, undefined structures (see Chapter 5). One possible cause of oiling out is that drops of the product solution are surrounded by the antisolvent, in which the solubility is very low, and this low solubility creates localized regions with very high supersaturation ratios. Before mixing to the molecular level is achieved, the localized high supersaturation forces the product out of solution without allowing sufficient time for the ordering of molecules to allow crystal development. The resulting oil particles have a tendency to clump together before the occluded solvent is dispersed throughout the bulk. As the mixture is aged, the oiled-out particles may transform into amorphous solids or become crystalline. Crystals developed in this manner will likely have a poorly defined structure. The problems that may result from such a scenario may include

- large drops of coalesced oil that will not disperse and can harden into a gum
- severe mixing issues due to the size and physical characteristics of the gum
- occlusion of impurities and solvent in the gum

The other potential issue is that the fine particles can stick together. The resulting agglomerates may be either strong enough to remain as agglomerates or weakly structured and reduce in size with continued mixing and/or subsequent processing (e.g., pump transfer). For a discussion of agglomeration, see Mersmann (2001), Söhnel and Garside (1992), and Chapter 5 in this book.

The above discussion of the potential negative effects from reverse addition (oiling out, poor crystal form, agglomeration) is necessarily qualitative, since this procedure, more so than many others, is strongly system dependent. The results could be satisfactory or unacceptable, depending on process requirements. Scale-up of reverse addition is particularly difficult because of the large deviations from equilibrium that are implicit in its implementation.

9.1.3 Addition Strategy

A common addition strategy is linear addition of the antisolvent. This addition strategy results in variable supersaturation as addition proceeds. As shown in Figs. 9-3 and 9-4 (curve A in both figures), supersaturation generally increases rapidly upon the addition of antisolvent at a constant rate. If the metastable region is as indicated, the supersaturation will exceed the limit very rapidly at the early phase of addition, and a nucleation-dominated process could result.

As shown in Figs. 9-3 and 9-4, curve B, the addition rate, can be slow initially and gradually increased to achieve a relatively constant level of supersaturation. The rate could follow a slowly increasing rate of addition, an analogy to the strategy of a programmed rate of temperature drop as originally proposed by Griffiths (1925) and further developed by Nyvlt (1971, pp. 118ff.). Other discussions of this control strategy may be found in Jones and Mullin (1974) and Myerson (2001, pp. 223ff.). In an ideal balanced situation with constant supersaturation and constant release of supersaturation via crystal growth and/or nucleation, the solution concentration profile will be maintained close to the solubility curve, as shown in Fig. 9-4 for programmed addition. This addition strategy has the potential to minimize nucleation and achieve growth by keeping the solution concentration within the metastable region. Growth potential is further enhanced by seeding, as discussed below.

9.1.4 Seeding

Seeding is addressed throughout this book, and many of the issues apply to antisolvent addition. The primary issue of when to seed is again critical to avoid excessive nucleation. The seed can be added as a powder or as a slurry in the antisolvent. The latter is preferred for

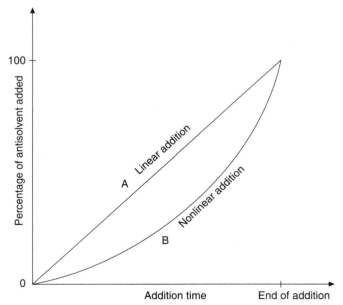

Figure 9-3 Addition rate profile for linear (curve A) and programmed (curve B) antisolvent addition rates. The linear addition profile adds antisolvent to the batch solution linearly over time. The programmed addition profile gradually increases the addition rate of antisolvent to the batch solution over time.

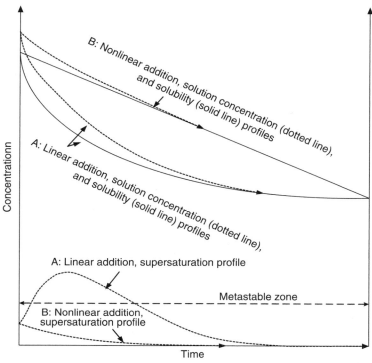

Figure 9-4   Solubility and supersaturation profiles for the linear and optimum addition rates. During the linear addition period, the supersaturation is not maintained at a constant level and will reach a maximum which may exceed the upper metastable zone boundary. For optimum addition, the supersaturation is maintained below a certain level and should approach the solubility limit.

many reasons, including ease of handling and satisfaction of containment requirements, as well as a possible increase in the effectiveness of the seed surfaces by "conditioning" in the antisolvent slurry to dissolve or remove shards.

Antisolvent addition strategies provide an alternative to the common method of addition of seed as a "shot." The shot method is subject to variable results because the seed may be subject to total dissolution if added too soon (before reaching point B, Fig. 9-1), or it may be ineffective in preventing excessive nucleation if added too late (after point B).

Since the seed is by definition not significantly soluble in the antisolvent, a slurry can be prepared of the seed in a small quantity ($\sim$5–10%) of the antisolvent, and this slurry is added to the substrate solution to achieve initial supersaturation. The timing of this addition is not as critical as that for adding a shot of seed, since it can be started near the saturation point and continued until the overall concentration is in the metastable region. The antisolvent in this slurry provides the driving force to reduce solubility and ultimately to reach saturation. Enough seed must be used since some of it will dissolve if the addition is started before the saturation point is reached. The difficulty is determining exactly where this point is from batch to batch, thereby presenting the same issue discussed for cooling and evaporative crystallization (Chapters 7 and 8, respectively). As in those cases, online analytical methods can be used to alleviate or remove this uncertainty. When these methods are difficult or impractical, the antisolvent slurry-seeding technique can be effective, and it has been successfully employed in several difficult situations.

Antisolvent addition processes are illustrated in this chapter and in Chapter 11.

EXAMPLE 9-1 *Crystallization of an Intermediate*

Goal: Improve the centrifugation and the drying rate of an intermediate

Issues: Slow centrifugation and drying caused by small mean particle size and bimodal PSD

In a multistep synthesis, plant productivity was limited by slow filtration and drying of a key intermediate. Analysis of the overall operation indicated that these problems resulted from the small mean particle size and bimodal PSD being obtained in the antisolvent addition step in the crystallization of the intermediate. Photomicrographs indicated some degree of agglomeration.

In this step, the antisolvent, water, was added to an initial solution of the intermediate in isopropyl alcohol/water over a 6 hour period, starting with a slow rate and increasing the rate as the addition continued, as discussed in Section 9.1.3 above.

The plant operation utilized a curved-blade turbine (CBT) impeller and a subsurface addition line that was suspected to be in a nonoptimum position and was too large in diameter. Presumably, the CBT was causing excessive shear and generating fines from the agglomerates. The position of the subsurface line, and its size, were suspected of promoting nucleation in the immediate region of the water stream inlet, an issue discussed in Section 6.4.4.1.

The first improvement in the plant was achieved by reducing the speed of the CBT impeller. A significant increase in mean particle size was achieved ($\sim$50 to $\sim$60 μm), although a bimodal PSD continued to be observed with fines in the 10–20 μm range. This distribution is shown in Fig. 9-5. Improvements in centrifugation and drying rate were achieved, but the fines content prevented a sufficient increase to achieve the desired productivity.

Simultaneously, a laboratory investigation was initiated to examine these effects in a scaled-down crystallizer designed to mimic the plant crystallizer as closely as possible. A pitched-blade turbine (PBT) was chosen to achieve good macromixing with reduced shear.

In the laboratory 1 liter crystallizer, a 1.5 mm subsurface addition line was positioned as close to the PBT as possible. The solution used in the lab crystallizer was taken from a plant batch, and the plant addition strategy was utilized.

A typical laboratory PSD result is shown in Fig. 9-6. Elimination of the fines by a combination of the change in the impeller and the subsurface addition line is apparent.

Figures 9-7 and 9-8 show typical crystalline product before and after the changes in impeller and inlet tube diameter, position, and orientation. The changes resulted in a $\sim$25% reduction in cycle time for centrifugation and drying.

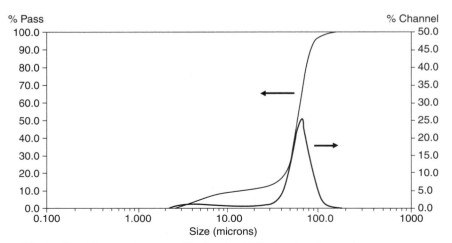

Figure 9-5   Profile of factory material crystallized with a CBT at low impeller speed.

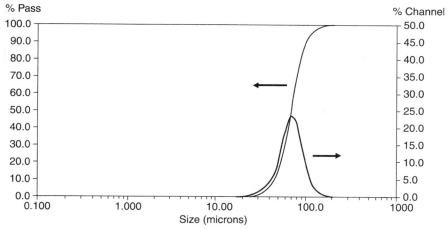

Figure 9-6   Profile of a lab run crystallized with a PBT, using an appropriately sized and positioned diptube.

Message

Anticipate changes in crystallizer conditions on scale-up. Although challenging, small-scale experiments can be made to mimic the poor mixing characteristics often encountered in larger-scale equipment.

EXAMPLE 9-2   *Rejection of Isomeric Impurities of Final Bulk Active Product*

Goal: Rejection of isomeric impurities to meet the purity specification

Issues: Coprecipitation of isomeric impurities, wash efficiency, occlusion

In the final step of the synthesis of an active pharmaceutical ingredient (API) up to 15% of diastereomeric impurities were present in the product solution after reaction and workup. Separation of one of the diastereomers ("SSR") from the desired isomer ("RRR") proved to be difficult and

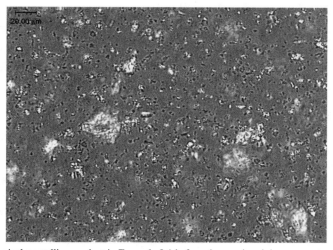

Figure 9-7   Typical crystalline product in Example 9.1 before changes in mixing parameters.

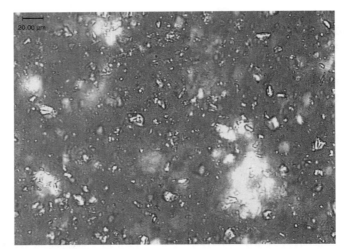

Figure 9-8 Typical crystalline product in Example 9.1 after changes in impeller and inlet tube diameter, position, and orientation.

inconsistent in the laboratory and pilot plant. Rework was required in order to reject the SSR impurity to below the specification of 0.8% in the final product. Figure 9-9 shows the structures of RRR and SSR.

OPTION 1

In the crystallization of RRR, the initial solution in ethanol is seeded (2%) and aged for 1 hour at 40°C. After aging, the slurry is cooled linearly to 20°C, followed by a linear addition of antisolvent (water) over 5 hours. A final cool-down to 0°C is performed to minimize the mother liquor loss. The final ethanol/water composition is 55/45.

The problem of rejection of one of the diastereomeric impurities, SSR, from the desired isomer, RRR, required an undesired rework in order to reduce the SSR impurity to below the specification of 0.5% in the final product.

OPTION 2

To understand the nature of SSR impurity in the RRR product, solubilities of SSR and RRR were measured. Figure 9-10 shows the solubilities of SSR and RRR in a neat ethanol/water solution and

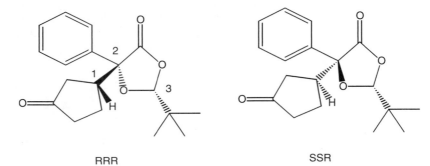

Figure 9-9 Molecular structures of RRR and SSR isomers. RRR and SSR have the same chemical structure and are different optical isomers.

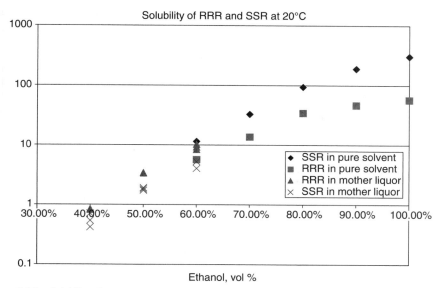

Figure 9-10   Solubility of RRR and SSR at 20°C in aqueous ethanol in pure solvent and mother liquors. As shown by the data, SSR is less soluble than RRR in mother liquor. SSR is more soluble than RRR in the pure solvent system.

in the actual mother liquor solution. The data indicate that SSR is more soluble than RRR in the pure solvent system but is less soluble in the mother liquor. While the cause of this reverse phenomenon is unclear, the data clearly show that SSR will not be rejected as favorably as was originally thought based upon the solubilities of SSR and RRR in the pure solvent system.

Additionally, supernatant samples were taken during crystallization and impurity profiles were tracked by high-performance liquid chromatography (HPLC) assays. As shown in Fig. 9-11, the RRR concentration was continuously reduced throughout the entire crystallization as expected. Interestingly, the SSR concentration remained constant up to the point of about 50% ethanol. Further reduction of the ethanol percentage caused a reduction of the SSR concentration. This was interpreted to mean that SSR was undersaturated and stayed in solution up to the point of about 50 vol% of ethanol. Beyond this composition, SSR crystallized out of the solution similarly to RRR. Removal of the impurity by washing of the cake after filtration was not consistently successful.

Other than the coprecipitation of SSR from the point of equilibrium solubility, SSR may also be occluded during crystallization. To evaluate this possibility, slurry and crystallization experiments were conducted side by side. The results are shown in Table 9-1. They suggest that occlusion may occur. For the case with 50% EtOH, cakes from the crystallization experiments gave a higher purity (94.45% and 93.78%) than the cake (91.66%) from the slurry experiment. However, for the case of 40% EtOH, there was no noticeable improvement of cake purity throughout crystallization. Nevertheless, since the specification of SSR in the final cake is below 0.5%, the improvement seen in the case of 50% ethanol simply cannot be ignored.

Based upon the solubility data in Figs. 9-11 and 9-12 and Table 9-1, several changes were made to optimize the crystallization. Table 9-2 shows the modified crystallization procedure.

With the modified procedure, the SSR level in the unwashed final product was effectively reduced to 0.3–0.4% consistently. After washing, the SSR level was less than 0.1%, which was below the specification.

Message

Coprecipitation of impurities can be experienced during crystallization when solubilities are exceeded. Solubilities may change in the presence of other impurities and may be reversed from those in pure solvent(s). Coprecipitation may also occur because of local conditions at the point of addition, leading

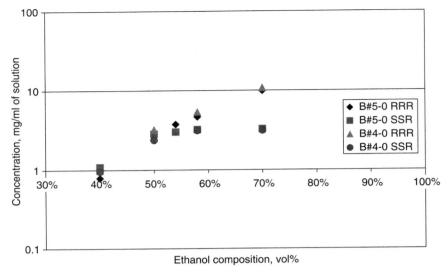

Figure 9-11 Supernatant concentration of RRR and SSR isomers at different solvent compositions. The data show that the SSR concentration remains constant until the solvent composition reaches approximate 50% ethanol. Below 50% ethanol, the SSR concentration drops linearly, similar to the RRR concentration profile. This suggests that SSR coprecipitates with RRR during crystallization below 50% ethanol.

to supersaturation of an impurity that can then cocrystallize and/or be occluded in the product; washing the wet cake may be ineffective in removing the impurities. Procedural modifications to control local supersaturation, including a change in composition of the antisolvent mixture, mode of addition, and location of the feed pipe, can be effective in preventing coprecipitation.

EXAMPLE 9-3 *Crystallization of a Pharmaceutical Product with Poor Nucleation and Growth Characteristics*

Goals: Improve particle size and filtration rate

Issues: Tendency to form oil or amorphous solid

Fine crystals in fibrous bundles

Laboratory crystallization of an API resulted in oil/amorphous solids or very fine crystals of high surface area ($>20\,m^2$/gm). Figure 9-12 shows microphotographs of the crystals. As a result of the fibrous nature of these crystals, the subsequent filtration is unacceptably slow. In order to improve the crystallinity and the filtration rate, an alternative procedure was developed.

Table 9-1 Conditions and Results of Slurry and Crystallization Experiments

Mode	Temp., °C	EtOH, vol%	Cake, HPLC area%		Mother liquor, HPLC area%	
			RRR	SSR	RRR	SSR
Starting material			83.3%	7.42%		
Slurry experiment,	20	40%	89.41%	5.71%	36.10%	18.57%
10 ml/gm of drug		50%	91.66%	4.22%	43.66%	22.59%
Fast crystallization,		40%	88.24%	6.05%	23.95%	15.29%
10 ml/gm of drug		50%	94.45%	3.05%	40.94%	22.86%
Slow crystallization,		40%	87.62%	6.58%	16.93%	10.35%
10 ml/gm of drug		50%	93.78%	3.58%	38.84%	20.59%

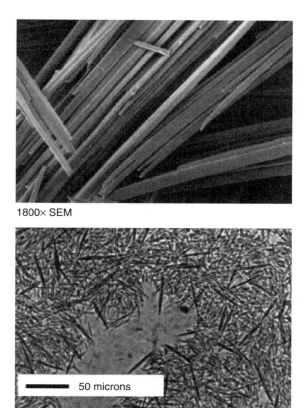

1800× SEM

50 microns

200× Optical

Figure 9-12   Scanning electron microscopic photographs of crystals for Example 9.3, Option 1, revealing the fibrous structure of the needle-like crystals.

OPTION 1

The procedure is illustrated in Fig. 9-13. A toluene solution of the compound is prepared and a small quantity of the antisolvent, acetonitrile, is added. The mixture is seeded and aged, and then the remaining amount of acetonitrile is added over several hours, followed by cooling and additional aging.

Table 9-2   Modification of Antisolvent Addition

	Original procedure	Modified procedure
Composition of antisolvent	100% water	Mixture of ethanol/water (to reduce local supersaturation at the addition point)
Addition time	Linear	Nonlinear (cubic) for constant release of supersaturation
Addition temperature	20°C	36–40°C to enhance crystal growth rate
Addition mode	Above surface	Subsurface, for better mixing and minimum local supersaturation
Final solvent ratio	55/45 ethanol/water	58/42 to avoid coprecipitation of SSR

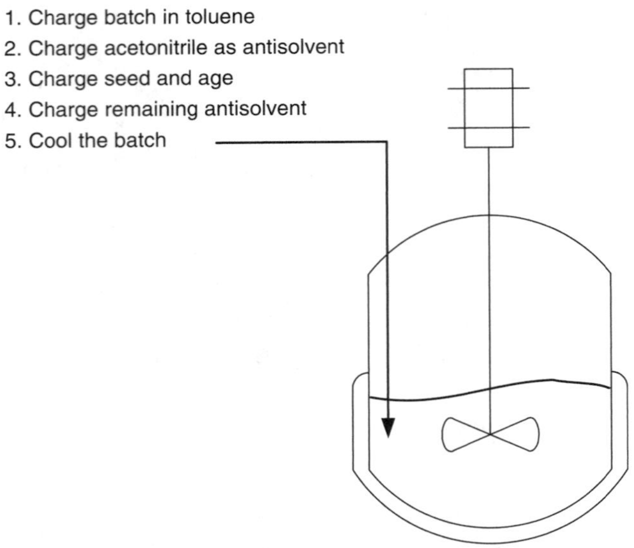

1. Charge batch in toluene
2. Charge acetonitrile as antisolvent
3. Charge seed and age
4. Charge remaining antisolvent
5. Cool the batch

Figure 9-13   Flow sheet of the antisolvent addition procedure for Example 9.3, Option 1. The antisolvent is charged to the batch, followed by cooling.

The objectives of this procedure were to create a mixture at low supersaturation and provide seed to prevent nucleation. After adding the seed and aging, the antisolvent was added slowly to minimize supersaturation, followed by cooling to reduce final solubility and maximize yield.

The difficulty in crystallizing this compound is further illustrated by the observation that if the antisolvent was added in less than 1 hour, the product would not nucleate at the resulting high supersaturation. Subsequently, oil/amorphous solids would form that would not become crystalline even with additional age.

The PSD under the best conditions with slow addition was still unacceptably broad and mean particle size was too small for efficient downstream operations in filtration, washing, and drying. The effect of PSD on these operations can be monitored very easily by measuring the filtration rate and evaluating the microphotographs.

OPTION 2

The procedure is illustrated in Fig. 9-14. A slurry of up to 10% seed crystals in the final solvent composition is charged to the crystallizer. A toluene solution of the compound is then added simultaneously with the antisolvent over a period of several hours at a constant temperature.

The filtration rate with the simultaneous addition procedure was 10 times faster than that of Option 1.

Message

The strong tendency of this compound to form oils or amorphous solids at increased supersaturation indicates a relatively narrow metastable zone width. Slow simultaneous addition with a large seed area effectively maintains the supersaturation sufficiently low to prevent nucleation of fines and allows some growth.

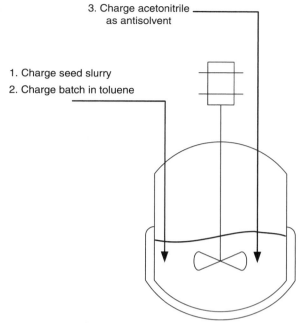

3. Charge acetonitrile
as antisolvent

1. Charge seed slurry
2. Charge batch in toluene

Figure 9-14   Antisolvent addition procedure for Example 9-3, Option 2. Both batch and antisolvent are charged simultaneously to the seed slurry.

OPTION 3

In addition to the development of the procedure outlined above, a factorial study was conducted in the laboratory to study several variables in an attempt to improve growth (Johnson et al. 1997; Togkalidou et al. 2001). The experiments consisted of a six-variable fractional factorial of 10 runs followed by experimental verification.

Input Variables

- mixing intensity (350–1450 RPM, 1.1–4.7 m/s tip speed)

- seed type (slurry or milled)

- seed level (3–10%)

- temperature (15–25°C),

- solvent ratio (50–70% antisolvent),

- addition time (1–6 hours)

Output/Response Variables

- filtration time and filtration cake resistance—using a $0.050\ m^2$ test filter with approximately 20 gm (40–50 mm cake height) of product at 20 PSIG filtration pressure

- Focused beam reflectance measurement (FBRM) particle size (chord length) distribution

- optical micrograph

Results from the factorial study indicate that the primary input variables affecting filtration resistance were seed type and mixing intensity. When slurry seed was used, insufficient slurry seed led to significant fines and a bimodal PSD, with the slowest filtration time. For milled seed, the large number of growth sites on the seed made the amount of seed the smallest contributor to filtration time.

Mixing sensitivity was found to affect the filtration time or filtration cake resistance. The higher the mixing intensity, the higher the filtration cake resistance. The solvent ratio was also an important factor affecting the filtration rate, followed by the temperature and addition time. A higher solvent/antisolvent ratio, higher temperature, and longer charging time all favor the reduction of cake filtration resistance.

For the output variables, as shown in Figs. 9-15 and 9-16, the filtration time data correlated well with mean particle size (chord length) and the level of fines by Lasentec FBRM measurement and optical micrographs. This correlation enabled direct feedback of process performance (cake filtration resistance) based upon FBRM measurement.

Based upon the results of the factorial study, the manufacturing-scale centrifuge filtration time (6000× scale) was reduced 73% from the simultaneous addition process developed in Option 2 above.

Message

The poor crystalline structure of this compound prevents good growth on seed even at low supersaturation. Mixing intensity is found to be a major variable in determining filterability. The slower the mixing, the faster the filtration time and the lower the cake filtration resistance.

EXAMPLE 9-4   *Impact of Solvent and Supersaturation on Particle Size and Crystal Form*

Goals: Control particle size via crystallization and avoid milling

Issues: solvate, generation of supersaturation, and selection of solvent

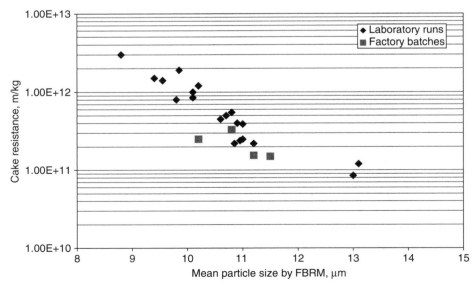

Figure 9-15   Effect of crystal size on filtration rate for Example 9.3, Option 3. The data indicate that the smaller the particle, the higher the cake resistance to filtration.

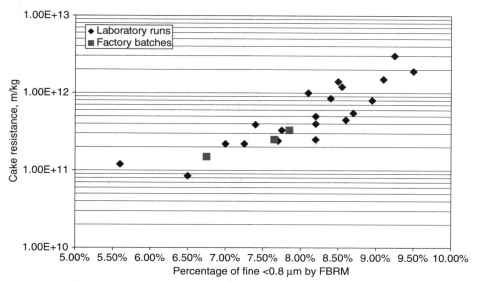

Figure 9-16 Effect of percentage of fines on filtration rate for Example 9.3, Option 3. The data indicate that the higher the percentage of fines in the PSD, the higher the cake resistance to filtration.

In the synthesis of a developmental API, dry milling was used to produce fine particles for formulation development. Due to concerns about solid handling during dry milling, development of a process to produce the desired particle size directly in the crystallization was initiated.

OPTION 1: NORMAL (FORWARD) ADDITION

The drug substance was dissolved in methanol at an elevated temperature. Isopropyl acetate as antisolvent was added. During addition of the antisolvent, spontaneous nucleation occurred. Antisolvent addition was followed by cooling to 0°C. A photomicrograph of the resulting crystal slurry is shown in Fig. 9-17. The final crystal size was in the range of 20–40 μm, thereby requiring milling to achieve the required mean particle size and PSD.

OPTION 2: REVERSE ADDITION

In an attempt to generate finer particles, a solution of the compound in methanol (MeOH) was added to the antisolvent isopropyl acetate (IPAC) in the reverse addition mode. The purpose was to generate a high degree of supersaturation in order to promote nucleation. As shown in Fig. 9-18, the primary particle size remained similar to that observed in Fig. 9-17. In addition, a higher degree of agglomeration was observed. Based upon these observations, it was concluded that this compound had very rapid crystal growth rate kinetics. Furthermore, rapid mixing would be required to avoid high local supersaturation, which can lead to agglomeration.

OPTION 3: IMPINGING JETS MIXING/SIMULTANEOUS ADDITION

Subsequent efforts focused on improving local mixing by using rapid mixing devices, i.e., impinging jets, without changing the solvent system. Interestingly, a new crystal form, which was identified as a methanol solvate, was generated (Fig. 9-19). The methanol solvate is more stable than the original

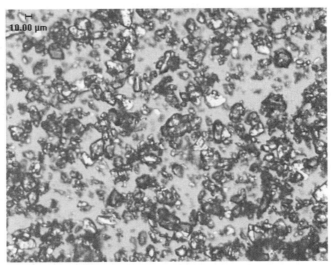

Figure 9-17   Microscopic photo of crystals for Example 9.4, Option 1 (normal addition). The cube-like crystals (significant third dimension) range from 20 to 40 μm in size, which is much larger than the milled material.

nonsolvated crystal form (Form A) in Option 1 above. However, during drying, the stable solvate converts to an unstable nonsolvate (Form B) which is less stable than the original nonsolvate (Form A). Unfortunately, Form B cannot be converted to Form A crystals in the drying step.

The presence of the new solvate created many complications for the development of the drug. It became impossible to crystallize the original nonsolvate crystals using the original MeOH/IPAC solvent system. Crystals generated by Options 1, 2, or 3 were all solvate. In order to avoid the complication of the solvate issue, the original solvent system, MeOH/IPAC, was abandoned.

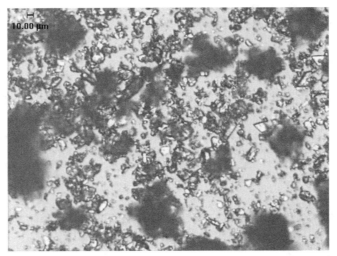

Figure 9-18   Microscopic photo of crystals for Example 9.4, Option 2 (reverse addition). The cube-like crystals range in size from 20 to 40 μm, with significant agglomeration.

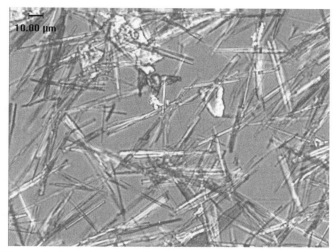

Figure 9-19   Microscopic photo of crystals for Example 9.4, Option 3 (impinging jet mixing/simultaneous addition). Needle-like crystals are formed under rapid mixing and high supersaturation instead of crystals with a third dimension. In addition, these needle crystals are a new solvate.

Solvent Selection

Various solvents and antisolvents were screened experimentally. Among them, dimethyl formamide (DMF) and IPAC appeared to offer the best combination based upon product solubility and the ability to retain the desired crystal form.

In comparison to the original solvent pair of MeOH/IPAC, the compound exhibited significantly different crystallization kinetics in the new DMF/IPAC system. Figure 9-20 shows the microscopic photo of the slurry sample when the antisolvent IPAC was charged to the solvent DMF (normal or forward addition). The particles were much finer in the DMF/IPAC system than those from the original MeOH/IPAC system. Fortuitously, the particle size was similar to that of the milled material and met

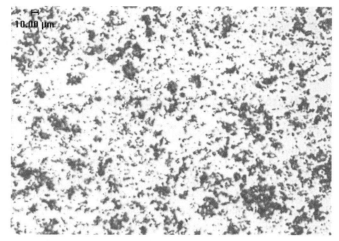

Figure 9-20   Microscopic photo of crystals for Example 9.4, Option 4 (normal addition, different solvent system). With a different solvent system, the crystals are much finer even in the normal addition operation, indicating significant impact of the solvent on crystallization kinetics.

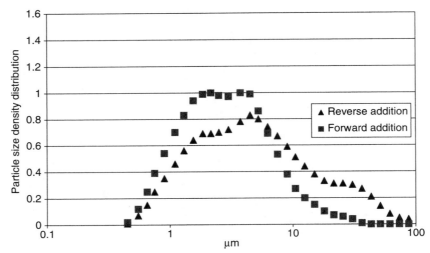

Figure 9-21 PSD of crystals for Example 9.4, Option 4 (forward vs reverse addition). Crystals formed under reverse addition shows a higher degree of agglomeration, as evidenced by the shoulder of the PSD.

the target PSD for delivery. As can be seen, in this particular case, the system had a very significant impact on the crystallization kinetics.

An additional study revealed similar agglomeration behavior when the batch in DMF was charged into the antisolvent IPAC solvent (reverse addition). As shown in Fig. 9-21, PSD data revealed distinct shoulders from 10 to 100 μm for reverse addition operation. Again, this result corroborated the criticality of local mixing for the reverse addition operation. Oil droplets were also observed during this mode of addition.

Since the DMF/IPAC solvent system with a forward addition mode gives a satisfactory PSD, it replaced the original MeOH/IPAC system. The process was demonstrated successfully in the subsequent pilot plant campaign.

Message

Solvent plays a critical role in the development of crystallization. It can affect strongly the final PSD and crystallization kinetics. However, a solvent is still selected primarily through empirical screening and may be dictated by a previous or subsequent step in the process. Also shown in this example is the impact of supersaturation on PSD and form (solvate, oil, and crystalline solid).

9.2 IN-LINE MIXING CRYSTALLIZATION

The focus of this book is on controlling crystallization processes at low supersaturation in order to minimize nucleation and thus promote conditions for growth. These processes are carried out primarily in stirred vessels, although in several examples fluidized beds are utilized. For some applications, i.e., pharmaceutical bulk products with low aqueous solubility, small mean particle size (3–5 μm or less and narrow PSD) are required for enhanced bioavailability or for inhalation therapy. In some cases, these particle sizes can be achieved by appropriate milling devices. However, milling can be undesirable because of environmental concerns about therapeutically active fine dusts. Installation and maintenance of the necessary engineering control and containment may be costly. In addition, milling can induce stresses and surface qualities such as electrostatic charges.

Development work for rapid mixing of reagents for fast chemical reactions has resulted in improved methods of contact. The preferred reactor design for extremely fast or mixing-sensitive reactions is an in-line mixer of appropriate design. Various possibilities have been investigated including centrifugal pumps (Bolzern and Bourne 1985); rotor-stator Mixers (Bourne and Garcia-Rosas 1984); impinging thin liquid sheets (Demyanovich and Bourne 1989); impinging jet mixing for reaction injection molding (Edwards 1984); and vortex mixing (Johnson 2003).

For crystallization, the objective of an in-line mixing device would be to mix the product solution and antisolvent to the molecular level in less than the induction time for nucleation. The impinging jet and vortex mixer designs can achieve these high degrees of micromixing at significant scale in an economical manner.

For impinging jet crystallization, industrial operation is described by Midler et al. (1994), with variants by Lindrud et al. (2001) (impinging jet crystallization with sonication) and by Am Ende et al. (2003) (specific reference to reactive crystallization). Laboratory studies are reported by Mahajan and Kirwan (1996), Benet et al. (1999), Condon (2001), and Hacherl (Condon 2003). Johnson (2003) and Johnson and Prud'homme (2003) report on the use of impinging jets to produce nanoparticles stabilized by block copolymers.

The local energy dissipation rate for an impinging jet can be $\sim 10\times$ that of a stirred vessel (see Paul et al. 2003, chapter 13). Impinging jet crystallization technology generates very high local supersaturation in order to create many fine nuclei. This process operates at concentrations—induced by antisolvent in the opposing jet—that are above the metastable upper limit, thereby inducing nucleation. Operation under this condition is feasible because control of nucleation can be achieved in a small, intensely mixed volume. This degree of control of nucleation is very difficult to achieve in a stirred vessel, despite application of current knowledge regarding localized distribution of energy dissipation.

Scale-up of an impinging jet process is simplified because the laboratory/pilot-scale feed line diameters may require only a two- or fourfold scale-up. The primary increase in scale is achieved by a longer run time. However, since the mixing device produces the same conditions of supersaturation at all times during operation, the time of the run can be increased manyfold without changing the nucleation conditions.

The small particle sizes that can be achieved by this technology can result in downstream operation issues with filtration rate, washing, and drying. These issues may be partially offset by the characteristic narrow PSD that can be effective in increasing the filtration rate. These issues must be considered in an overall assessment of the use of this technology compared to the potential advantage of eliminating milling (other than delumping).

EXAMPLE 9-5   Crystallization of an API Using Impinging Jets

Goal: Crystallize product requiring a high specific surface area, good crystallinity, and high chemical purity

Issues: Controllable, reproducible specific surface area for product crystals. High crystallinity to meet storage life requirements. Rejection of impurities.

The product being crystallized was subject to a familiar set of constraints in the pharmaceutical industry. In early clinical trials, it was established that for bioavailability the active drug had to be supplied with a high specific surface area ($2.3-4.0 \text{ m}^2/\text{gm}$). On the other hand, it had to be highly crystalline, since (accelerated) stability testing had shown that partially (or totally) amorphous product, often encountered in the production of small (high surface area) particles, was subject to unacceptable degradation in storage. To make matters worse, uneven cake wash had to be avoided because of the presence of an unacceptable impurity in the mother liquors.

Process development for this process followed a familiar scenario for production of high bioavailability product.

OPTION 1: PRODUCTION OF LARGE CRYSTALS, FOLLOWED BY MILLING

This option incorporates the growth-driven crystallization described as desirable throughout this book, with a milling process that negates some of the positive features, by

- creating particles with stress fractures, increasing the rate of decomposition in storage;
- producing, in general, a broader size distribution (more fines and large nuggets) than would be created by a crystallization step alone, even with delumping;
- generating the possibility of amorphous content in the batch at points of localized high impact energy and temperature buildup; and
- adding a potentially noisy and dusty operation in the plant.

Some pharmaceutical products are enough of a health hazard to preclude dry milling as a processing step. High-energy wet milling might be considered for these compounds, although the above negative considerations still hold, and problematic filtration and drying steps might still be encountered.

OPTION 2: NUCLEATION-DOMINATED CRYSTALLIZATION WITH ANTISOLVENT ADDITION IN A HIGH-ENERGY DISSIPATION LOCATION IN A STIRRED VESSEL

As discussed in Chapter 6, high-energy dissipation zones have been identified for certain stirred tank/impeller configurations. These zones are often small, and they can move enough so that the exact location of and linear velocity from an addition dip pipe are very difficult to optimize. When *very* intense micro- and/or mesomixing are required, stirred tanks are not the ideal type of equipment to carry out a robust, reproducible process.

For the crystallization process in Example 9.5, variation of supersaturation in a stirred tank nucleation process could not simultaneously meet the surface area and purity specifications for the product.

OPTION 3: NUCLEATION-DOMINATED CRYSTALLIZATION BY ADDITION OF SOLUTION TO ANTISOLVENT (REVERSE ADDITION)

As discussed in Section 9.1.2, reverse addition has been a time-honored method of producing small crystals. It typically suffers from broad size distribution and high impurity content (if present in the mother liquors) because of the excessive supersaturation gradients encountered by the inlet solution. In the present example, the final product contained too much of an undesired impurity. This product will also be compared to that from the impinging jet operation in the discussion below.

OPTION 4: NUCLEATION-DOMINATED CRYSTALLIZATION WITH SIMULTANEOUS SOLUTION AND ANTISOLVENT ADDITION IN A SEMICONTINUOUS IMPINGING JET APPARATUS

Impinging jet crystallization was discussed in Section 9.11 above. One configuration of this type of operation is shown in Fig. 9-22. The impinging jet contacting device delivers its product to an age vessel (either batch or continued stirred tank CSTR) to provide an age time, required for most compounds, to allow diffusion of mother liquors from the "droplets" (actually nucleated solids with trapped

High Surface Area Crystallization

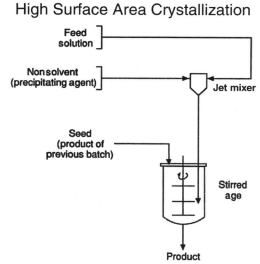

Figure 9-22   A common flow diagram for impinging jet crystallization.

solvent) and amorphous-looking solids initially nucleated. In very short residence time CSTR experiments, the required time to "age" the solid into material showing a highly crystalline structure was about 5 minutes. Aging time for other compounds has varied from seconds to minutes.

Seed addition to the age vessel (or in the antisolvent stream) is not necessary to attain small crystalline particles, but in this case seed was found to improve the reproducibility of the cycle time for the process.

The impinging jet apparatus for any given scale of operation is small and highly productive. Figure 9-23 shows a small laboratory version which could produce 6 kg/hr of the compound in this example. Actual lab experiments were run for 1 minute to produce ~100 gm of product. The impinging jets can be operated confined, as shown, or open (immersed in the age tank). In the latter case, the localized kinetic energy in the jets is, in practical terms, not adversely affected by the macromixing in the vessel. Despite the convenience of the open jet operation, there are advantages to a confined operation, of the type noted above in Fig. 9-22. Johnson (2003) studied the effect of confined impinging jet geometry on performance.

As an alternative to impinging jets, experiments were conducted in which solution and antisolvent were contacted in a Y mixer arrangement. Product from this arrangement had properties similar to those of product previously produced by reverse addition. The liquids contacted each other, and nucleated, in a planar interface with a very sharp supersaturation gradient. Clearly, the micro- and mesomixing effects created by the impinging jets are needed to break down these gradients prior to nucleation.

Jet Diameter: 1 mm
Productivity: 6 kilos / hour

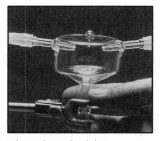

Figure 9-23   Impinging jet crystallization has very high volumetric productivity.

Processes occurring in parts of the system shown in Fig. 9-22 are as follows:

Jet Mixer (1.4 Seconds in Lab Operation)

- Mixing of the liquids
- Nucleation
  - Primary
  - Secondary
- Exclusion of impurities
- Amorphous to crystalline conversion (some materials)
- Crystal growth (some materials).

Transfer Line (0.2 Second in Lab Operation)

- Nucleation
- Crystal growth

Age Vessel

- (Solubilization of amorphous?)
- Secondary nucleation
- Crystal growth
- Exclusion of impurities
- Ostwald ripening (fines dissolution)

Product from the impinging jet operation had excellent X-ray diffraction patterns, equal in peak intensity to the most highly crystalline large particles crystallized in a growth-dominated process.

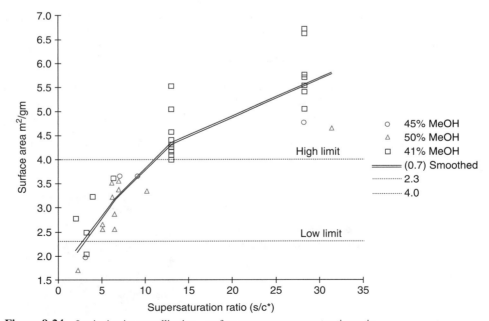

Figure 9-24   Impinging jet crystallization—surface area versus supersaturation ratio.

Particle size (surface area) reproducibility was excellent in the desired range. Higher temperature was also found to improve reproducibility. Since the desired specific surface area range was 2.3 to $4.0\,\mathrm{m^2/gm}$, a condition was chosen to produce $3.0\,\mathrm{m^2/gm}$ product. A series of six identical crystallization experiments had a mean of $3.05 \pm 0.4\,\mathrm{m^2/gm}$ (0.4 was the standard deviation for the set).

Figure 9-24 shows that the specific surface area of the crystallized solid was a function of the supersaturation ratio. That figure also shows that operation aiming for a higher surface area $(5-6\,\mathrm{m^2/gm})$ was subject to considerably greater variation than that in the desired $2.3-4.0\,\mathrm{m^2/gm}$ range.

Impurities rejection for solid crystallized with the impinging jets was good at 25°C and excellent at 50°C. For an undesired contaminant required to be <0.5 wt%, the impinging jet operation at 50°C

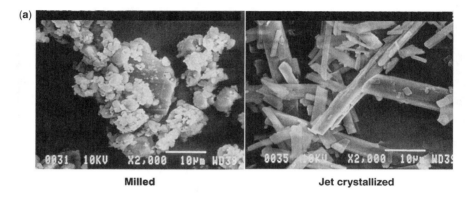

(a)

Milled   Jet crystallized

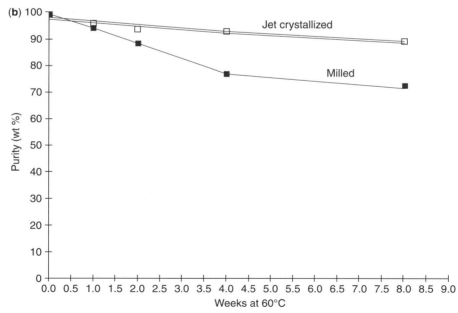

Figure 9-25   (a) Photomicrographs of equal surface area products obtained by milling larger crystals and by impinging jet crystallization. (b) Improved product stability of high surface area crystalline product from impinging jet crystallization.

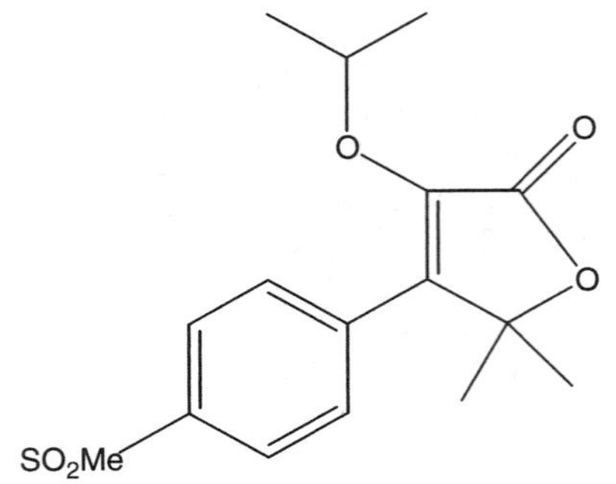

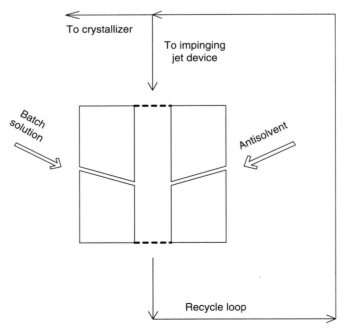

DFP

Figure 9-26   Chemical structure of DFP, an API candidate utilizing impinging jet crystallization technology.

consistently made product with <0.1 wt% of that impurity, comparable to that produced by slow growth-dominated crystallization. In contrast, impinging jet crystallization carried out at 25°C created product with 0.2–0.4 wt% of that compound (still acceptable), whereas reverse addition crystallization had 0.7 wt% (unacceptable).

Milled versus Jet Crystallized Final Product

Figure 9-25a shows electron micrographs of milled and impinging jet crystallized product. Both have the same specific surface area. The superior size uniformity of the jet crystallized material is clear. The

To crystallizer

To impinging
jet device

Batch
solution

Antisolvent

Recycle loop

Figure 9-27   Schematic of the impinging jet crystallizer showing the option to recycle around the mixing device. The recycle loop provides additional turbulence for the exit stream after the impingement.

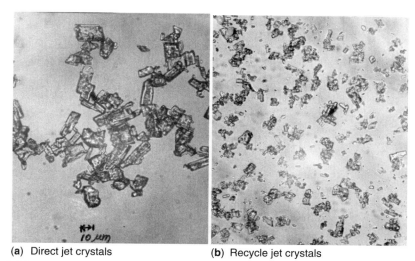

(**a**) Direct jet crystals   (**b**) Recycle jet crystals

Figure 9-28   Microscopic photos of (a) direct jet and (b) recycle jet crystals.

milled product obtains its surface area from very small particles produced in the mill, despite the fact that there are still some relatively large particles in the population.

Product Stability

Figure 9-25b shows the results of an accelerated (60°C) stability test for solid crystallized via impinging jets, versus that for milled product with the same specific surface area. As discussed earlier in this example, the milled crystal surfaces contain many stresses. Stressed surfaces on a solid are more prone to decomposition, thereby reducing its stability, as would be expected.

Table 9-3   Impact of Solvent Ratio, Temperature, and Concentration on Product Surface area

Run no.	Solvent ratio (Acetone/H_2O)	Overall conc. (mg/ml)	Supersaturation (overall conc./ solubility)	Temp. (°C)	Surface area (m^2/gm)
150-1	1/3	50	116	25	n/a
150-2	1/3.7	53	355	10	1
190-1	1/3.8	31	65	40	0.6
190-2	1/4	50	139	25	0.9
190-3	1/6.3	34	156	25	1
200-1	2/7/1 (ethanol)	50	63	25	0.5
227	1/3	63	481	8	0.9
234	1/5	42	463	5	1.2
241	1/5	42	174	25	0.9
248	1/4	50	417	15	1
PPB 1-6	1/4	46	n/a	10	1.3

Message

Control of fluid dynamics can assist crystallization processes in certain circumstances, and it is incumbent on the development chemists and engineers to fully understand the sensitivity of the process to these parameters.

<table>
<tr><td>**EXAMPLE 9-6**</td><td>*Crystallization of a Pharmaceutical Product Candidate Using an Impinging Jet with Recycle*</td></tr>
</table>

Goals: To achieve rapid mixing via an impinging jet in order to generate uniform fine crystals

Issues: Solubility, solvent composition, temperature, crystal surface area, bioavailability

As shown in Fig. 9-26, a developmental drug candidate, "DFP," was chosen to study this technology. This compound has low solubility in water, making particle size and surface area important factors in determining its bioavailability.

Figure 9-27 shows the recycle-loop impinging jet device used in this study. The normal mode of impinging jet mixing operation (one case being that in Example 9.5) does not incorporate a recycle loop. However, one can be used. For this particular case, the recycle stream provided additional turbulence in the exit stream after the impingement, creating product with a smaller mean size and a narrower size distribution, as shown in Figs. 9-28a and 9-28b.

Since the recycle stream contains crystals, this approach relies inherently on the secondary nucleation phenomenon under high supersaturation. This is the key difference between the normal mode and recycle mode of operation. Nucleation occurs in the loop, and the slurry in the loop returns to the crystallizer (Tung et al. 2008).

Selection of solvent/antisolvent, temperature, and overall concentration were critical parameters in affecting the supersaturation as well as the nucleation and crystal growth rates. Table 9-3 summarizes the experimental conditions and results of impinging jet crystallization of DFP. Figure 9-29 further shows the final product surface area as a function of supersaturation. As shown in Fig. 9-29, higher

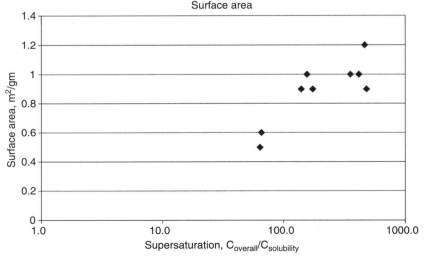

Figure 9-29   Surface area as a function of supersaturation. The higher the supersaturation, the higher the surface area.

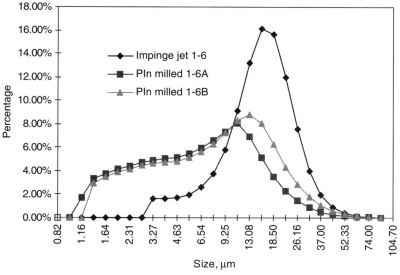

Figure 9-30   PSD of the same compound (DFP) crystallized by impinging jet compared to that crystallized in a stirred vessel followed by pin milling. Impinging jet crystals have a narrower PSD than the milled material.

Table 9-4   "DFP" Plasma Level

	Rats		Dogs	
Batch	AUC (μg $\times$ h/ml)	C$_{max}$ (mg/ml)	AUC (μg $\times$ h/ml)	C$_{max}$ (mg/ml)
16 (pin milled)	40.2 ± 5.5	4.8 ± 0.8	25.3 ± 6.1	4.5 ± 1.0
150-2 (impinging jet)	46.1 ± 2.8	4.9 ± 0.4	37.6 ± 5.6	4.7 ± 0.2
190-1 (impinging jet)	43.3 ± 7.3	4.7 ± 1.0	38.3 ± 9.6	4.5 ± 0.6

supersaturation resulted in a higher surface area. Higher ratios of antisolvent to solvent, lower temperatures, and higher concentrations all led to higher supersaturation. These parameters were used to tune the PSD and surface area accordingly.

A side-by-side comparison of PSD between the milled material and the impinging jet crystals is shown in Fig. 9-30. It is clear that the milled material exhibits a wider PSD with a higher percentage of fines than the impinging jet crystals. Upon handling, the impinging jet crystals had better solid flow properties and appeared to be less fluffy, tacky, and electrostatic than those from the milling operation.

The bioavailability data are shown in Table 9-4, where AUC stands for "area under the curve." As shown in Table 9-4, there is no significant difference in the bioavailability of the milled and impinging jet crystals.

Message

Impinging jet technology is capable of achieving a small mean particle size with a narrow PSD. Several parameters are significant in determining the final PSD, including solvent ratio, temperature, and concentration. These parameters affect supersaturation, which in turn affects particle size and surface area. A recycle mode of operation was used in this particular example. Higher supersaturation generated finer particles with a higher surface area.

Chapter 10

Reactive Crystallization

10.1 INTRODUCTION

When supersaturation of a crystallizing compound is created by its formation by chemical reaction, the operation is termed reactive crystallization. These operations are also known as precipitation. The term reactive crystallization is generally applied only when the product is crystalline. The products of the more general term precipitation may be amorphous or crystalline.

The reaction may be between two complex organic compounds or may be neutralization by an acid or a base to form a salt of a complex compound. These reactions can be very fast compared to both the mass transfer rates to the crystals and the growth rate of the crystals, possibly leading to high local supersaturations and, therefore, extensive nucleation.

Although crystallization by antisolvent addition shares many characteristics with that caused by chemical reaction, the processes often differ in the rate of creation of supersaturation (e.g., a rapid reaction leading to a compound of very low solubility). Reactive crystallization is also subject to other kinetic considerations which are sometimes less predictable than the known solubility effects caused by addition of an antisolvent.

In some cases, the fine crystals or precipitates resulting from high supersaturation (often in the range from 0.1 to 10 μm) are desired in order to meet specific needs for downstream processing or formulation. In most cases, however, these fine particles are not desired since they can be very difficult to handle in downstream processing—notably filtration, washing, and drying.

The reader will note many uses of qualitative terms to predict the behavior of these complex systems. As in the entire field of crystallization, these wide brackets around possibilities (e.g., will it crystallize, will it form an oil first, will it stay amorphous, will it grow, will it nucleate, what is good mixing, what is low supersaturation, etc.?) are necessary because of the extreme species and conditions dependency of the crystallization of organic molecules. The guidelines offered are intended as such and, in addition, to provide a framework for experimentation to determine where a particular system may fit in the wide scope of crystallization behavior possibilities.

10.1.1 Utilization

This method of crystallization has become increasingly common in the pharmaceutical industry because organic molecules often have poor water solubility and must be converted to a salt form to improve water solubility and enhance bioavailability. Reactive

Crystallization of Organic Compounds: An Industrial Perspective. By H.-H. Tung, E. L. Paul, M. Midler, and J. A. McCauley
Copyright © 2009 John Wiley & Sons, Inc.

crystallization/precipitation is also used in the fine chemicals industry to create fine particles for a wide variety of applications including photographic chemicals, dyes, printing inks, agrochemicals, topical formulations, and cosmetics.

The ultimate particle size distribution (PSD) from crystallization is dictated by the balance between nucleation and crystal growth rates. The processes indicated above, because of very high supersaturation, often result in rapid nucleation of too many particles and smaller than desired final product. The rapid nucleation and growth occurring may also result in impurity and/or solvent occlusion.

In addition, these normally small particles may agglomerate into weak structures which are readily broken up during aging in the crystallizer, transfer to a filter, on the filter, and/or in the dryer. The resulting low filtration rates, high solvent retention, poor washing, caking, and slow drying may be impractical for a production operation. Viscosities of the product slurries may be high because of high solids contents and poor two-dimensional crystals. In addition, other properties of the bulk active product such as bulk density (high bulk), PSD (excessively fine, broad, and/or bimodal), and caking may be unsuitable for pharmaceutical formulation operations.

Intermediates in a chemical synthesis may also be produced by reactive crystallization and are also subject to impurity occlusion and poor downstream performance.

10.1.2   Literature

The literature contains some excellent discussions of the theory and practice of reactive crystallization. In particular, the reader is referred to the definitive book on this topic by Söhnel and Garside (1992), which contains both theoretical development of the key factors involved in precipitation and a summary of practical aspects.

In addition, excellent treatment of this topic may be found in the books by Mullin (2001) and Mersmann (2001). The major part of this literature is focused on the study of precipitation of inorganic compounds by reactions between ions. Specialized studies of organic compounds are also treated including, for example, ethylenediaminetetraacetic acid (EDTA) (Myerson et al. 2001) and calcium oxalate (Marcant and David 1991).

Mixing issues and chemical reactions are the subject of definitive books including Baldyga and Bourne (1999) and Zlokarnik (2001).

In addition, several studies have appeared concerning theoretical and experimental studies on the effect of mixing on precipitation. These include Manth et al. (1996), Houcine et al. (1997), Torbacke and Rasmuson (2001), Åslund and Rasmuson (1992), and Phillips et al. (1999).

In their chapter in Mersmann (2001), Klein and David summarize the difficulties in scale-up as follows: "The classical agitated industrial vessel seems to be too complicated in terms of turbulence or macro- and micromixing to allow a reliable scale-up from laboratory experiments" (p. 557).

These scale-up issues present some of the more difficult challenges in crystallization processes, as they do for other engineering operations. In most cases, the products obtained may be handled but with increased capital and operating expense for the cumbersome downstream operations. These costs may exceed those of the other steps in a manufacturing operation. In extreme cases, the products may be unsuitable for commercial use because of the poor physical properties discussed above.

This chapter outlines some guidelines that may be helpful in some cases in developing process options that are capable of improving the chemical purity and/or physical properties of the product of a reactive crystallization. These guidelines are discussed in the text to follow and illustrated in Examples 10-1 and 10-2. It is recognized that there are systems for which these alternative process options will not be successful in increasing particle size or improving purity. The system characteristics that cause the most difficulty are short nucleation induction time, rapid nucleation rate, and slow growth rate. The last of these can cause the most difficulty and, in the extreme, some compounds in which nucleation represents a substantial fraction of the release of supersaturation do not grow beyond the nucleation/growth phase. However, it is recommended that a development program include investigation of the options before concluding that no improvement in the basic particle obtained from a reaction can be realized. Guidelines for development are presented.

10.2 CONTROL OF PARTICLE SIZE

Because the rate of creation of supersaturation in reactive crystallization is dependent on the reaction kinetics, control of particle size can be difficult because slowing down the reaction is often difficult or undesirable. For compounds with fast crystallization kinetics, traditional methods such as reduced concentration and temperature may help, but the range of improvement may not be significant.

The rate of addition of the reagents, however, does provide a means to control supersaturation globally in the reactor but not locally since the reaction may be complete near the point of addition. Successful operation depends, therefore, on a careful balance between addition rate of the reagent(s), local supersaturation, global supersaturation, mass transfer, and crystal growth surface area. For compounds with slow crystallization kinetics, the rate of addition of reagents provides an effective means to control supersaturation.

Three possible courses of a reactive crystallization are illustrated in Fig. 10-1. The equilibrium solubility line is essentially horizontal since most reactive crystallizations result in

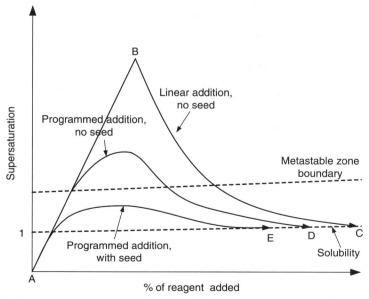

Figure 10-1 Schematic representation of addition modes for reagents in reactive crystallization: the relationship between the amount of reagent added and the metastable region with and without seed.

products that have low solubility in the solvent system and, therefore, may have small changes in solubility based on temperature. Because of the high supersaturations generated, primary nucleation is usually dominant, producing a large number of crystals. Three reagent addition strategies are illustrated; linear (A, B, C), programmed (A, D), and programmed with seed (A, E). It should be noted that the solubility at a supersaturation of 1.0 will normally be very low and the first amount of reagent added will create relatively high local supersaturation.

The differences in crystallization conditions for these three additions are shown by the degree of the addition that occurs above the metastable region, thereby producing excessive nucleation in the unseeded cases. This excessive nucleation may be avoided by addition with a sufficient amount and size of seed to maintain supersaturation within the metastable region. This strategy is discussed in Section 10.3.5.

Controlled supersaturation at the initiation of the addition of the reagent(s) requires an initial charge of seed to minimize uncontrolled nucleation and the resulting creation of an excess number of particles. It is desirable to develop the seed in a separate operation because it is difficult to ensure a robust process with a reproducible seed level using only the intrinsic reaction. This issue and methods are discussed in Section 10.3.4 below.

10.2.1 Controlling for Growth

The objective of the remainder of this chapter is to suggest methods of achieving growth-dominated processes of organic compounds by reactive crystallization. Two examples of controlled processes are presented (Examples 10-1 and 10-2). In these cases, successful scale-up was achieved by minimizing nucleation and promoting growth, thereby helping to minimize the scale-up issues. As also indicated, this is not to suggest that all reactive crystallization processes will be responsive to the same methods since this approach requires that the compound in question has reasonable inherent growth rates. In the case of amorphous products, these process options are not expected to be of value because predictable and consistent solid phase growth rates are very difficult to obtain with amorphous solids. A possible exception is "growth" by agglomeration which is controlled by other factors. Reliance on agglomeration may be feasible in some cases and is discussed in Section 10.3.8 below.

The creation of fine particles may be desired for specific down-stream processing or formulation requirements. Methods for control of these applications are also discussed in Section 10.5 below.

The complex issues regarding low solubility and the nucleation characteristics of induction time and nucleation rate must be minimized by the maintenance of low supersaturation so that growth can predominate. Methods for promoting this balance will be indicated. Prior to that discussion, the issues encountered in the development of a reactive crystallization process will be briefly reviewed.

10.3 KEY ISSUES IN ORGANIC REACTIVE CRYSTALLIZATION

Although the references cited above are focused primarily on inorganic precipitation, the key issues are essentially the same for organic compounds, with the exception that nucleation and growth can be slower in many cases. However, since many organic bases and acids are formed by reactive crystallization from their respective parent forms, the reactions can be very fast, leading to high supersaturation with potential for nucleation, as in inorganic precipitation.

As is elaborated in all of these discussions, the final physical attributes and possibly the chemical purity may be a function of several rate processes that are occurring both in series and in parallel. The most common procedure is to add a reagent to a solution of the organic compound to form the crystallizing species. The resulting rate processes include

- mesomixing and macromixing of the added reagent to a degree sufficient to start a molecular-scale contact and reaction
- micromixing to molecular-scale homogeneity combined with mesomixing
- the chemical reaction itself to produce the reaction product at low supersaturation
- the formation of clusters of the reaction product in solution, ultimately exceeding the solubility
- the creation of an insoluble phase consisting of the reaction product now as an amorphous "oil" or solid or crystalline solid—nucleation
- continuation of supersaturation as the reagent is added
- continuation of simultaneous nucleation and growth
- possible secondary changes including agglomeration, flocculation, crystal breakup by shear, change in crystal form, ripening
- completion of addition
- continuation of secondary changes

Note: Precipitates that are initially an oil or amorphous can "turn over" and become crystalline at any point after formation of clusters or remain amorphous.

These many interacting rate processes may all be at play in a reactive crystallization. Successful development of a consistent and scalable process requires evaluation of the potential for each to be significant so that methods of control can be developed.

Two additional factors that have potential for critical influences are

- impurities—sometimes referred to as additives if they are intentionally introduced
- mixing both in a mixer configuration and at the addition point in a vessel

The critical issues that must be addressed in experimental development and scale-up studies are outlined below. Mixing issues are involved in all aspects and will be discussed first.

10.3.1 Mixing Issues

The extensive literature on the key mixing issues of macromixing, mesomixing, and micromixing, for chemical reactions is applicable to mixing issues encountered in reactive crystallization. A major focus of this literature is on the effect of mixing on complex reactions with multiple products without separation of a second phase. The issues in reactive crystallization are usually confined to reactions with a single product that does separate as a second phase. The reader is referred to Chapter 6 in this book and the references cited therein for a review of this field of mixing. The following is a brief discussion of mixing issues in reactive crystallization.

10.3.1.1 *Micromixing*

Micromixing is a key factor in reactive crystallization because the time of blending to the molecular level is critical to both the chemical reaction and the induction time for nucleation. Micromixing times in stirred vessels are critically affected by

- location of the reagent(s) feed pipe(s)
- impeller speed
- impeller type

The range of micromixing times at various locations in a stirred vessel can vary by a factor of >20 at the same impeller speed and also by large factors for different speeds and types. This time is critical, depending on the reaction rate and the nucleation induction time, in determining the nucleate particle size and the complex sequence of events in the micromixed zone. For fast reactions and short nucleation and induction times, a particle leaving this zone may already be a crystalline or an amorphous solid. For slower reactions and longer induction times, no phase change may occur in this zone.

A critical factor in micromixing and mesomixing in stirred tanks is the location of the reagent feed pipe(s). Experiments at different feed pipe locations with the reactive crystallization under investigation can readily determine whether micromixing and/or mesomixing issues are controlling. A change in mean particle size and PSD at different feed pipe locations would indicate mixing sensitivity. On the other hand, if no significant differences are observed, a different crystallization property of the system, such as nucleation induction time, may be slow relative to the micromixing time. The reaction itself could also be slow and occur in the bulk mixing regions of the vessel. Experimental work on this effect for barium sulfate is described by Nienow and Inoue (1993) and for calcium oxalate by Marcant and David (1991).

Organic compounds exhibit nucleation induction times over a wide range. Faster times are on the order of milliseconds that are in the same range as micromixing times in stirred tanks. Mixing effects can then be expected to be significant on particle formation. Reagent feed into poorly mixed regions can experience 10 to 100 times slower micromixing and mesomixing times, thereby making mixing issues more severe even with longer nucleation induction times. Reaction times can bracket these time frames, thereby adding to the complexity of particle formation. These effects are illustrated by the Damkoehler number, as discussed in Chapter 6 and illustrated in Fig. 6-2.

Micromixing times can be reduced by an order of magnitude by in-line mixers such as impinging jets, as discussed in Chapter 9. These devices can be utilized effectively in reactive crystallization and are discussed in Ph.D. theses on this topic by Condon (2001) and Johnson (2003), in journal articles by Johnson and Prud'homme (2003) and Hacherl (Condon) (2003), and in Section 10.5 below.

10.3.1.2 *Macromixing*

Macromixing is another term for bulk blending in stirred vessels. Compounds with longer nucleation induction times and/or reagents with slower reaction rates may not form particles in the micromixing or mesomixing zones but will eventually do so in the bulk blending phase of the overall mixing process. Although these are reactive crystallizations, they will be influenced more by anti-solvent crystallization effects of mixing.

10.3.1.3 *Mesomixing*

Mesomixing effects occur in stirred vessels when the reagent(s) feed rate is faster than the local mixing rate, resulting in a plume of reagent concentration that is not yet mixed to

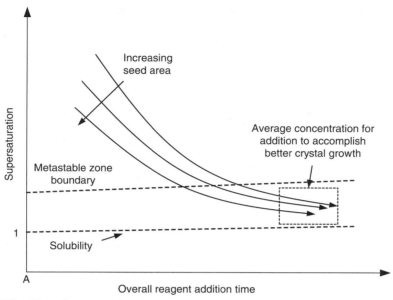

Figure 10-2   Schematic representation of reagent addition time (overall time for a run) and the metastable region indicating the recommended region for promoting growth and minimizing nucleation.

the molecular level. This mixing effect is important on scale-up and results in the need for longer reagent addition times on the larger scale to achieve the same reaction selectivity as on a smaller scale (Balgyga and Bourne 1999, pp. 732–744). Its effect on reactive crystallization would be expected to require increased addition time to achieve an equivalent mean particle size and PSD, although data on this dependency are difficult to document because of the many interacting variables.

10.3.2 Mixing and Growth

Mixing effects are also critical in reactive crystallizations designed to minimize nucleation and maximize crystal growth. This design strategy is discussed in Section 10.3.6 below. The issue here is to achieve micromixing of two reagents faster than the nucleation induction time and reaction time as well as effective meso- and macromixing in order to create supersaturation at a controlled rate. Under these conditions and in the presence of sufficient seed surface area and mass transfer to the growing crystals, nucleation can be minimized in favor of growth. However, excess mixing could cause a decrease in induction time and an increase in nucleation rate as well as crystal fracture by shear, thereby requiring a balance in power input.

These contrasting mixing requirements form the basis of the mixing dilemma that is particularly applicable to reaction crystallization. It is apparent that each reactive system will respond differently because of the extreme ranges in organic compound inherent characteristics including reaction rate, induction time, nucleation rate, growth rate, and shear sensitivity. These issues are discussed in the references cited above and in Chapter 6. Laboratory experimentation is required to find the balance required. Scale-up studies are then necessary to ensure that these balances can be maintained at the predicted conditions. Examples of a growth-dominated reactive crystallization are presented below (Examples 10-1 and 10-2).

10.3.3   Induction Time and Nucleation

Induction time, nucleation rate, and nucleate particle size all can play key roles in determining the course of a reactive crystallization. These issues are discussed in the crystallization literature and in Chapters 2, 4, and 5 in this book for general crystallization, and their importance in reactive crystallization will be summarized here.

Induction time is a key issue because of its relationship to reaction time and mixing time. At the point of reagent introduction, as outlined above, mixing, reaction, and nucleation may be occurring within the same time frame more or less in parallel or may occur in series, depending on the respective time constants. If they occur in parallel, mixing may not be complete to the molecular level before reaction starts. If induction time is also short, incorporation and/or entrapment of unreacted substrate in a nucleating mass in regions of local high supersaturation are possible. In this case,

$$t_{mix} \sim t_r \sim t_{ind}$$

Ideally, the mixing is sufficiently intense locally to achieve reagent blending to the molecular level before reaction, and the reaction is complete in less than the induction time so that nuclei can form from the expected molecular composition. In this case,

$$t_{mix} < t_r < t_{ind}$$

Mixing time can be varied as discussed above, but reaction and induction time are both system specific. Reaction time and induction time are both concentration dependent, and induction time also is a function of supersaturation (t_{ind} decreases with increasing S).

Also because of high supersaturation, nucleation may be primarily homogeneous rather than secondary, although this is also system dependent and may be reversed for organic molecules. Nucleation rate is a key issue because of its impact on the number of crystals formed.

These interacting factors may be uncontrollable without control of local conditions to prevent extensive nucleation.

10.3.4   Supersaturation Control

Control of both local (point of addition) and global supersaturation is essential, as in all crystallizations, if a satisfactory balance between nucleation and growth is to be achieved. This is particularly relevant to reactive crystallization because of the creation of local high supersaturation of these low-solubility compounds that is unavoidable at the point of reaction. In addition to the mixing issues outlined above, the critical variables in minimizing supersaturation are as follows:

1. Addition time of the reagent—must be long enough to maintain the global supersaturation in the metastable region.
2. Sufficient initial seed area—to avoid or minimize nucleation at the outset of addition and prevent formation of an excessive number of nuclei, which would limit overall growth.
3. Continuing balance between addition rate and growth surface area—to promote growth and avoid continuing nucleation, which would result in a bimodal distribution.

These issues are shown schematically in Figs. 10-1 and 10-2. Figure 10-1 is discussed in Section 10.2 above. Figure 10-2 shows a similar relationship, and the region of concentration in the metastable zone to be maintained throughout the addition is indicated.

(a)

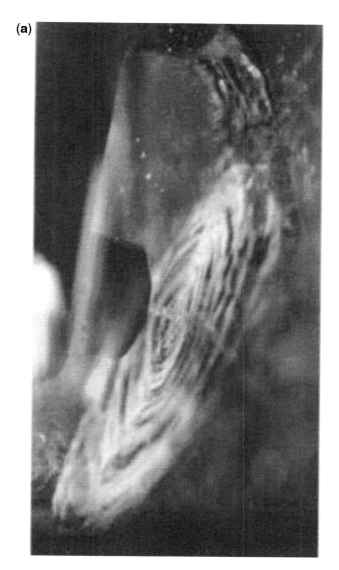

Figure 10-17 (a) Photographs of impingement planes in a free jet at feed velocities of 5.1 and 5.5 msec of calcium chloride and sodium oxalate, respectively. The intensity of mixing is clearly shown.

(b)

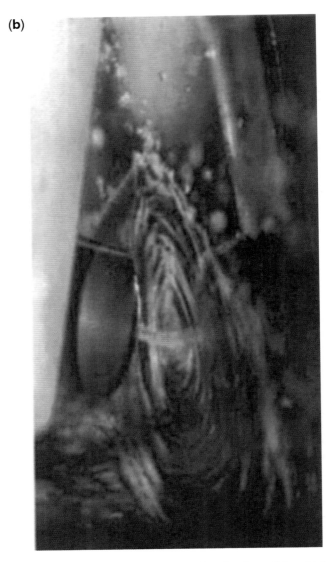

Figure 10-17 (b) Photographs of impingement planes in a free jet at 5.5 and 5.1 msec of calcium chloride and sodium oxalate, respectively. The intensity of mixing is clearly shown.

10.3.5 Seeding

Seeding is the key to achieving control of a reactive crystallization process. Without seeding, excessive nucleation can be expected in most systems, resulting in severe limitation on the final crystal size by creating an excessive number of particles. The guidelines outlined below have been found to be useful in several cases and will be further illustrated in Examples 10-1 and 10-2. For low growth rate compounds, however, growth of suitable seed material and significant growth of crystals during reagent addition may not be realized. In this case, a growth-dominant process may not be achievable.

1. The number, surface area, and surface condition of seed crystals are critical to successful minimization of nucleation and realization of growth. (Seeding issues are also discussed in Chapter 5, and only those issues of primary consideration in reactive crystallization are discussed here.)

2. The recommended seed mean particle size is approximately one-half of the expected final mean particle size.

3. The recommended seed amount is at least of 10–15% (the amount necessary to maximize growth by providing sufficient initial surface area to minimize nucleation).

4. The recommended method to add seed is in a slurry or as a heel from a previous batch. Drying and milling are not recommended. Wet milling or sonication is recommended between batches to maintain the desired seed size as necessary.

5. Recommended seed preparation is by recrystallization from a suitable solvent(s). (Initial preparation by reactive crystallization without seed will typically give seeds that are too small for significant growth in successive use.)

Note: Seed growth by heat/cool temperature cycling is an effective means of growing seeds for initial use. This method is also useful in determining the potential for a particular compound to grow. This method is discussed in Chapter 5 and in Example 10-1 below.

The requirement for increased amounts of seed for reactive crystallization compared with the cooling, evaporation, and anti-solvent methods is discussed by Mullin (2001, p. 339). Amounts of seed up to 50% are indicated to be necessary in recycle systems "to provide the seed area necessary." The requirement for this increased amount is the direct result of the rapid development of supersaturation by reaction and the need to have sufficient surface area for growth throughout the operation, especially at the start of reagent addition.

10.3.6 Crystal Growth

The first issue with regard to growth is to determine if the compound in question has potential for growth. Growth potential may be very difficult to determine definitively because of (1) large dependence on impurities and (2) lack of patience. Once established, growth can be used for initial seed preparation. Once seed is grown, all crystallizations are started with proper seeds.

Agglomeration and/or aggregation are common in reactive crystallization and should not be confused with true growth (see Section 10.3.8.1 below). Secondary growth phenomena can also be expected, such as growth rate dispersion and size-dependent growth (see Section 10.3.8.2 below).

10.3.6.1 Addition Strategies for Growth

To increase the probability of achieving and maintaining a primarily growth process, the entire operation must be maintained in the metastable region, as shown schematically in

Figs. 10-1 and 10-2. The three critical factors that can achieve this condition are (1) mixing to minimize local supersaturation at the point of addition, (2) limiting the addition rate to prevent buildup of a bulk concentration of the reaction product, and (3) sufficient seed surface area. The rates of mass transport through the film surrounding each crystal and the incorporation into the crystal lattice by surface integration reach a balance so that they are essentially equal. The key to successful operation is to maintain the bulk concentration sufficiently low to allow the rate of surface integration to control so that transport through the film does not create a region of high concentration at the surface that could result in local nucleation. The effects of addition time and seed area on growth conditions are illustrated in Fig. 10-2.

Since the inherent growth rate of many organic compounds is relatively slow, addition times may be long in order to achieve supersaturation control within the metastable zone. Higher addition rates can result in nucleation and the creation of a bimodal distribution. Experimentation to determine acceptable addition rates can be evaluated by focused beam reflectance measurement (FBRM) and other in-situ, online methods (Chapter 2) or microscopic observation of the crystal slurry which could reveal the presence of fines. These issues are highlighted in the examples below.

An addition strategy, termed programmed feed concentration, has been proposed by Lindberg and Rasmuson (2000) in which the initial feed concentration of the reagent is kept low and then increased as the addition proceeds. This strategy is analogous to programmed cooling (Chapter 7) or to programmed addition of antisolvent (Chapter 9).

By limiting the reagent concentration at the most critical stage, the resulting mean particle size can be increased.

A method of determining the maximum addition rate consistent with avoiding nucleation is to experiment with increasing feed rates with all other variables held constant. The addition rate at which nucleation is observed, as indicated by the appearance of a bimodal distribution, is then the limiting addition rate. In-situ methods of making this determination are valuable in this experimentation, as discussed in Chapter 2.

10.3.7 Impurities/Additives

The large dependence of growth rate on the presence or absence of impurities is well known and can be critical to the success of achieving growth of seed and product. Growth can be totally inhibited by levels of impurities as low as 0.1% but be satisfactory at >50%. This extreme variation is a function of the molecular structure of the impurity and its ability to retard or block surface incorporation. Excellent discussions of these issues may be found in the referenced texts. Söhnel and Garside (1992, p. 159) note that despite significant study, "no fully reliable general theory that does not need experimental verification of each case has so far been worked out to allow prediction of the influence of a certain impurity on the behavior of a given system or of the concentration at which the admixture becomes effective." A case of dissolved impurities causing significant decreases in growth—including stopped growth under supersaturated conditions—may be found in Example 11-6.

Initial attempts to establish growth potential must be accompanied by careful evaluation of impurity contamination. This can be difficult when the structure of the impurities may not be known. One experimental technique to evaluate the effect of impurities that are inherent in the process but whose structures may not be known is to recrystallize the subject compound both from pure solvents and from the same solvents "spiked" with mother liquors from a primary crystallization. This technique can be very successful in revealing whether crystal growth is affected by the impurities in the spiked liquors.

In any event, extensive purification in the laboratory by any applicable method, including preparative high-performance liquid chromatography, is recommended as part of growth potential determination studies. The importance of this effort cannot be over-stated since a particular compound *may not grow* only because of the presence of impurities and not because of inherent crystal lattice incorporation restrictions.

The use of additives to influence crystal habit has received much experimental and theoretical attention. However, it is not discussed here because these techniques are not generally utilized in the pharmaceutical industry due to limitations on the addition of additives.

10.3.8 Secondary Effects

10.3.8.1 Oiling Out, Amorphous Solids, Nucleation, Agglomeration, and Growth

Reactive crystallization operations are subject to oiling out and/or agglomeration because of the inherently high local supersaturations encountered. As indicated in Section 10.3, the formation of a crystal may be preceded by oiling out as the first physical form that may or may not be observed (see also Chapter 5, Section 5.4). This oil may separate as a second phase because of the normally extremely low solubilities of the reaction products that result from the chemical reaction. This low solubility can cause a second liquid phase to form on a time scale that is shorter than the nucleation induction time. These issues are considered in Ostwald's Rule of Stages.

The next event could be solidification of the oil into an amorphous solid or direct nucleation to a crystalline form. In either case, the fine particles are subject to agglomeration as the particles collide and stick together because of the oily/sticky nature of the solid surfaces at this point. Assuming a crystalline form results from this series of events, these nuclei can then grow to some extent, as determined by the system-specific characteristics of the compound. This process continues throughout the addition of the reagent(s).

The final mean particle size and PSD will be determined by the number of nuclei created by these steps. If nucleation continues throughout the addition, the mean particle size will be small, perhaps $< 10 \, \mu m$, and the PSD will be broad and possibly bimodal because of the large number of nuclei produced and the conditions allowing some crystals to grow while nucleation is creating small crystals.

If, on the other hand, there is a sufficient quantity of seed present initially and if the reagent(s) is added at a sufficiently slow rate with sufficiently good mixing, the low supersaturation created could be well below the metastable limit, thereby reducing or eliminating nucleation and supporting growth on the seed crystals. As indicated above, 10–20% of seed crystals could then be expected to approximately double in size, depending on the aspect ratio of the crystal.

The oiling out and agglomeration possibilities can complicate or prevent successful crystallization. In an extreme case of oiling out, a solid phase never forms and the oil drops could coalesce or remain dispersed. This condition could result from several factors, including a high impurity level inhibiting nucleation, a low melting point solid, and further depression of the melting point by mutual solubility of the solid and a component of the solvent mixture.

Agglomeration can complicate an operation by forming large structures of crystals or amorphous solids that can trap impurities, unreacted reagents, and/or solvents. These structures may be strong enough to stay coalesced and persist through filtration, washing, and drying. In some cases, agglomeration can be beneficial to these downstream operations if the trapping of impurities is not excessive.

In many cases, however, agglomerates are weakly structured and break down on continued mixing, during pumping, during filtration, and/or during drying. This disintegration could be difficult to handle in these downstream operations, and primarily for this and impurity entrapment issues, agglomeration should be avoided when possible.

Methods to avoid oiling out and agglomeration are essentially the same as those described above for favoring growth. The single most important consideration is avoiding high local supersaturation to the greatest extent possible by good mixing, seeding, and slow addition.

10.3.8.2   *Growth Rate Dispersion and Size-Dependent Growth*

Other secondary effects, which are not exclusive to reactive crystallization, are size-dependent growth and growth rate dispersion. These effects may not be separable, but both can change the final mean particle size and PSD. Both are discussed in Söhnel and Garside (1992, pp. 103–105) and in Chapter 4 of this book.

10.3.8.3   *Reactive Crystallization to Increase Reaction Selectivity*

The selectivity of a consecutive chemical reaction may be increased above that predicted from its rate constant ratio if the reaction is run under conditions in which the product crystallizes while the reaction is in progress. Crystallization of the product reduces its concentration in solution, thereby making it less available for over-reaction. The topic of increasing reaction selectivity in the presence of, or by the creation of, a second phase (e.g., by crystallization), has been discussed by Sharma (1988) and Paul et al. (2003, chapter 13). This important application of crystallization is illustrated in Examples 11-2 and 11-3.

10.4   SCALE-UP

Development and scale-up of reactive crystallization/precipitation processes can present some of the most difficult challenges in the field. The reader is referred to the quotation in Section 10.1.2 above as a reminder of this difficulty.

However, the authors have participated in development and scale-up of some successful reactive crystallization processes, and the examples to follow (Examples 10-1 and 10-2) are included to illustrate the concepts and application of the principles discussed above in these processes. These developments were based on the three essential concepts of seeding, control of supersaturation and promotion of growth, as described above. The key variables are, therefore,

- amount of seed
- size of seed
- addition rate
- mixing intensity

EXAMPLE 10-1   *Reactive Crystallization of an API*

Goals: Development and scale-up of a robust process for the reactive crystallization of an API suitable for downstream formulation

Issues: Crystals formed by reaction are thin needles with an aspect ratio of >10.1 and a diameter of ~2 μm—unsuitable for downstream processing or formulation

INITIAL PROCESS

The process for the final step in a multistep synthesis produces a sulfate salt by the addition of an ethanolic-sulfuric acid solution to an ethanolic solution of a free base. The resulting sulfate salt crystals are shown in Fig. 10-3a.

These needles are unsuitable for downstream filtration, washing, and drying as well as for blending and filling of capsules because of several characteristics of this type of needle, including low bulk density (~ 0.08 ml/gm) as dried product. Changes in the procedure were necessary to solve the severe problems that would result from attempted scale-up to manufacturing at the required high throughput.

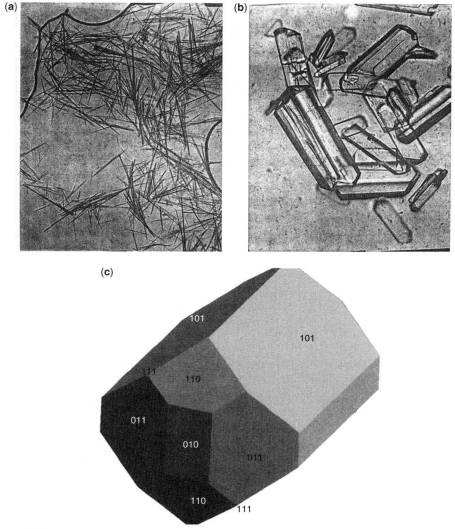

Figure 10-3 (a) Needles from the reactive crystallization of an API as described in Example 10-1. The needles are unacceptable for this API and result from overly rapid addition with insufficient seed. (b) Crystals of the API described in Example 10-1 after growth from the needles shown in Fig. 10-3a. (c) Crystals of the API described in Example 10-1 as predicted from crystallographic model.

As discussed above, the causes of these fine crystals include

- high supersaturation at the point of mixing of the acid reagent with the basic organic substrate
- uncontrolled nucleation
- generation of an excess number of particles
- limited potential for growth on the excess numbers of particles

POSSIBLE SOLUTIONS

There are two commonly used methods:

- a slower addition rate to reduce supersaturation globally in the vessel
- increased mixing to reduce local supersaturation at the point of contact of the reagents

Neither of these commonly utilized methods proved to be satisfactory in this case because the reaction is still too fast for local control, and increased mixing—even if effective—would break the fragile needles.

The Dilemma

These problems are characteristic of reactive crystallization and are commonly handled by the provision of downstream equipment to cope with the poor filtration and washing characteristics, the high solvent content of the filter cake, and the resulting slow drying and possible caking. These plant operations are both capital and labor intensive.

In this example, these problems are exacerbated by the high production requirements and by the formulation difficulties of filling capsules with low bulk density needles. Therefore, even if the production issues could be tolerated, the formulation problems could not.

REVISED PROCESS

A method to resolve these issues is based on revisiting the basic issues of reactive crystallization as summarized in Section 10.3, beginning with the amount and size of seed. These seeding issues are key to providing sufficient surface area for possible growth on the seed to relieve the local supersaturation before excessive nucleation occurs. Seeding possibilities and their limitations using needles produced by the best conditions above are as follows:

- 1–5%, insufficient surface area for growth at the high supersaturation
- >5%, excess number of fine particles; even if no additional nucleation occurs, the needles can only double in diameter—still too small.

Neither of these seed amounts is satisfactory since the lower amount will result in excess nucleation and the higher amount will result in only a limited increase in the size of the basic needles.

Since no amount of fine needle seed is satisfactory, the next possibility is to increase the particle size of the seed. To accomplish this, it is necessary to grow seeds in a separate operation from the reactive crystallization.

Growth of Seed

In the separate operation, starting with the fine needles, these crystals are subjected to many heat/cool treatments with sonication between cycles. A suitable solvent is required in which the solubility approximately doubles on heating (e.g., a temperature increase of 20–50°C). During heating, the finer crystals will dissolve. During slow cooling, growth on the remaining crystals may be achieved. Sonication after cooling can break the needles lengthwise. In subsequent cooling cycles, the diameter of the needles, the slowest growth dimension, can slowly increase, eventually producing three-dimensional crystals.

The needles shown in Fig. 10-3a were subjected to this heat/cool treatment, resulting in the large three-dimensional crystals shown in Fig. 10-3b, in remarkable agreement with the prediction from crystallographic modeling (Fig. 10-3c). Success in growing seed crystals both prepares of seed for larger batches and establishes that this compound will actually grow. It is then necessary to determine if it will grow at a practical rate in the actual reactive crystallization system.

Amount of Seed, Mixing, and Addition Rate

Using these large seed crystals, runs of increasing size were made in the laboratory to make enough seed for a pilot plant batch. The amount required, as indicated above, is 10–20% to ensure sufficient surface area for growth.

A sufficiently slow addition rate was found, as determined by decreasing this rate until a condition was found that did not result in a bimodal distribution with fine needles. Such formation would indicate that the addition rate was too fast or there was insufficient surface area for growth, thereby resulting in local high supersatuation and nucleation. A bimodal distribution is shown in Fig. 10-4.

Optimization of mixing is also critical because of the need to achieve rapid blending while avoiding crystal fracture.

Continuous Process

The high production rate required for this API was compatible with continuous operation in a stirred vessel. This option was tested at the 0.004 m³ (4 liter) scale. The initial operation was satisfactory. However, after several hours, fine needles started to appear, as shown in Fig. 10-5. The cause of

Figure 10-4   Crystals from Example 10-1 showing a bimodal distribution during crystallization because of insufficient seed growth area.

Figure 10-5 Crystals from Example 10-1 showing the effect of continuous crystallization, with decreasing seed area for growth causing formation of fine needles.

fine needle formation was decreasing seed surface area resulting from all-growth and decreasing particle count from harvesting of product.

Although methods to avoid this could have been implemented, this option was not pursued because of development time pressures. It is discussed here as an illustration of the key importance of surface area. This factor in a continuous crystallization process is also discussed in Example 11-6.

Final Process

Satisfactory crystal size and shape are achieved with the following combination of these factors:

- 15–20% large seed
- 2 hour addition time
- good blending without excessive shear (as provided by an Ekato Intermig impeller)
- wet milling between batches to maintain the mean particle size and PSD of the seed within suitable limits

Scale-up was successful in producing crystals, as shown in Fig. 10-6. Mixing was based on equal power per unit volume. A sub-surface addition line was provided for the acid addition. The bulk density increased from 0.08 to 0.35 ml/gm.

Message

This example demonstrates that a reactive crystallization process can be developed to produce significantly larger crystals than are naturally formed in most cases. Successful development requires

Figure 10-6   Final crystals from the manufacturing scale, Example 10-1.

definition of a combination of the four critical factors that can allow growth to be predominant over nucleation.

In some cases, successful growth may not be achievable because of an intrinsically short induction time, high nucleation rate, fast reaction relative to these factors, and/or slow or no growth potential. In the event that growth cannot be observed, purification to remove possible growth inhibitors may be effective in promoting growth and leading to successful optimization of the four key variables.

This example applies to only true crystallization systems. If the solid product is an amorphous solid, growth by agglomeration may be possible but true crystal growth is not.

EXAMPLE 10-2   *Reactive Crystallization of an Intermediate*

Goals: Development of a robust process to produce a filterable intermediate with high purity

Issues: Laboratory operation produced bimodal distributions with unacceptable filtration rates and impurity entrapment

INITIAL PROCESS

The laboratory process to produce an intermediate in a multistep process involves reaction of two organic compounds that produce a crystalline product as the reaction proceeds and is, therefore, a reactive crystallization. The initial process produced crystalline product that was irregular in shape and difficult to filter. In addition, the chemical purity was severely reduced by contamination with one of the reactants.

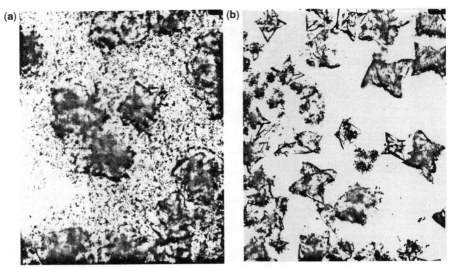

Figure 10-7 Crystals from Example 10-2. (a) Bimodal distribution from the original process showing fines formation caused by fast addition time and (b) crystals from a manufacturing-scale operation without contamination with fines.

The reaction was carried out by the addition of reagent B to A over the course of 1 hour. In this mode of operation, contamination of the product with reagent A was unacceptable at 4–5% and filtration rates were low (0.36 m^3/hr-m^2).

Reduction in the amount of A in the product was investigated first by varying the addition scheme. Simultaneous addition of A and B was effective by minimizing the amount of unreacted A in the mixture at all times. A reduction of A contamination was achieved. However, the filtration rate was still unacceptable for scale-up.

As in Example 10-1, the issue was addressed by seeding and a reduction in the feed rates of the reagents. Seed did not have to be grown separately, as the problem in this example was bimodal distribution and not initial size. The bimodal distribution is shown in Fig. 10-7a.

FINAL PROCESS

The final process utilizes simultaneous addition of A and B over a 5 hour period to a seed bed from the previous batch (~15%). The impeller selected was the Ekato Intermig for the same reasons as in Example 10-1.

The product from the manufacturing-scale operation is shown in Fig. 10-7b. The prevention of fines formation resulted in an increase in filtration rate of 5×. A low reagent A contamination of 1% was also realized. Using these conditions, scale-up from laboratory to pilot plant to manufacturing, overall 2000×, was successful. The effect of addition time on the amount of the reaction/crystallization actually carried out in the metastable region is shown in Fig. 10-8.

Message

As with any crystallization process, reactive crystallization will, in general, produce fine particles unless the entire operation is run within the metastable region. This condition can be realized by provision of heavy seeding and by slow addition to control supersaturation at a low level. Adequate mixing is necessary, but shear damage must be avoided by selection of the correct impeller speed and type.

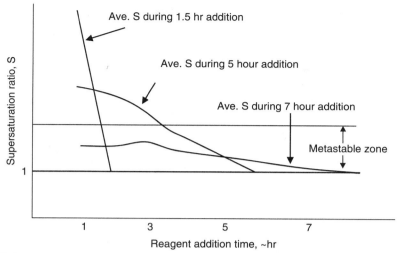

Figure 10-8   Reagent addition time effect on the supersaturation experienced during addition.

Instrumentation

The development programs outlined in Examples 10-1 and 10-2 would have benefited from the in-line instrumentation process analytics technologies (PAT), as discussed in Chapter 2, that is now available but was not when these processes were developed.

EXAMPLE 10-3   *Reactive Crystallization of a Sodium Salt of an API*

Goals: Investigate mixing, addition rate, and seed impact on the residual solvent of a sodium salt of free phenol API

Issues: Crystals formed by reactive crystallization contain residual solvent up to 0.5 wt%, which is the limit of specification for release of bulk drug.

OPTION 1: ORIGINAL PROCESS

The process for the final step in a multistep synthesis produces a sodium salt of free phenol. Figure 10-9 shows the flowsheet of the crystallization process. Two streams—stream A containing free phenol dissolved in *n*-methyl pyrolidone (NMP) and stream B containing ethanolic sodium ethoxide solution—are charged simultaneously to a slurry seed bed of sodium-salt of phenol. The addition time is several hours. At the end of each batch, a portion of the batch slurry is wet-milled and used as seed for the next batch. The amount of seed varies from 5% to 20% of the size of the batch.

Figure 10-10 shows the photomicrograph of the crystals. Some layering of the crystals, indicating fast crystal growth, is apparent. The resulting product contains varying levels of residual NMP up to 0.5 wt%, which is the limit of residual solvent for this compound. Therefore, it is critical to understand the underlying crystallization mechanism in order to resolve this issue.

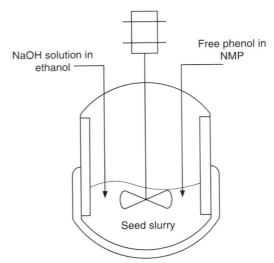

Figure 10-9   Flowchart for the simultaneous addition of stream A containing free phenol in an NMP solution and stream B containing sodium ethoxide in ethanol for Example 9-3.

OPTION 2: REVISED PROCESS

As discussed in Chapter 2, solvent can be occluded in the crystal lattice due to fast crystal growth and/or excessive nucleation. To test this hypothesis, a series of experiments were conducted. The following parameters were varied:

- seed type: milled versus unmilled seed
- seed loading: 5% to 15%
- addition time: 30 minutes to 8 hours
- solvent ratio of EtOH/NMP: 50/50 and 30/70

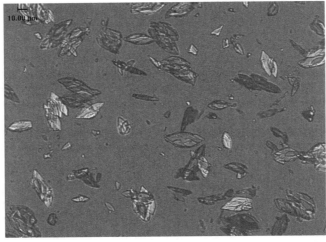

Figure 10-10   Microscopic photo of crystals in Example 10-3. Some layering of crystals is apparent, indicating rapid crystal growth.

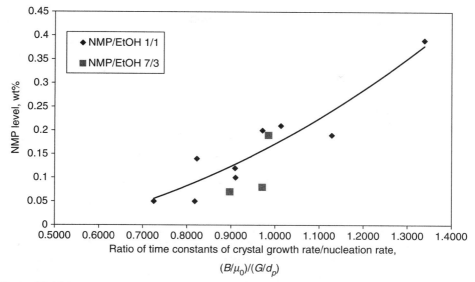

Figure 10-11   Residual solvent in the cake as a function of the ratio of time constants of crystal growth rate to nucleation rate. The higher the ratio, which means more rapid nucleation than crystal growth, the higher the residual solvent in the cake.

The PSD during the crystallization was monitored by a Lasentec FBRM in-situ particle size analyzer. Figure 10-11 is a plot of the residual solvent in the cake versus the ratio of time constants of crystal growth rate over nucleation rate. The time constant for crystal growth rate, $1/(G/d_p)$, is defined as the inverse of crystal growth rate over mean particle size averaged over the entire period of crystallization. The time constant for nucleation rate, $1/(B/\mu_0)$, is defined as the inverse of nucleation rate over the total number of particles averaged over the entire period of crystallization. Both time constants can be calculated directly from the FBRM (for simplicity, the procedures used to calculate the growth rate and nucleation rate are not shown here).

Here, the attempt is to apply these simplified time constants to reflect the relative rates of crystal growth and nucleation during crystallization. A high ratio of crystal growth time constant over nucleation time constant means that it takes more time for crystal growth and less time for nucleation. Hence, there will be a higher degree of nucleation and a lower degree of crystal growth in the process. Similarly, a low ratio of crystal growth time constant over nucleation time constant means that it takes less time for crystal growth and more time for nucleation. Hence, there will be more crystal growth in the process.

As shown in Fig. 10-11, the occluded residual solvent was found to be a strong function of time constants for crystal growth rate/nucleation rate over the entire range of parameters studied. Specifically, it was shown that the higher the ratio of $(B/\mu_0)/(G/d_p)$, which means more rapid nucleation rate, the higher the residual solvent, and vice versa. This finding clearly shows the importance of properly controlling the nucleation rate and the crystal growth rate in the process.

Recommended Process

After identifying the criticality of controlling the nucleation and crystal growth rates, the process is revised, with particular attention given to seed size, addition time, and mixing configuration. Table 10-1 shows the results of residual solvent in the cake with seed type, additional time, and mixing mode as the operating parameters.

From Table 10-1, it appears that as long as any two of three factors—seed type (fine-milled material vs coarse-milled material), addition time (4 vs 8 hours), and mixing configuration (above

Table 10-1   Impact of Experimental Conditions on Residual Solvent in the Cake

Experimental conditions			Results
Seed type	Addition time, hr	Mixing mode	Residual solvent in the cake, wt%
Fine milled material by air-attrition mill	8	Above surface	0.09–0.14
	4	Above surface	0.24
	4	Rapid mixing device	0.12–0.15%
Coarse-milled material by a rotor-stator mill	8	Above surface	0.22

Note: 15 wt% seed was used in all experiments.

surface vs rapid mixing device)—are chosen properly, the residual solvent in the cake can be controlled below 0.2 wt% as required. For example, using "fine-milled material by air-attrition mill" and "8 hours addition time" will produce a cake with 0.09–0.14 wt% residual solvent, even though the streams were charged "above surface." However, using "fine-milled material by air-attrition mill" but "4 hours addition time" and "above surface" addition produces a cake with a level of 0.24 wt% residual solvent. Unfortunately, the experiment using "fine-milled material," "8 hours addition," and "rapid mixing" was not tested due to the cancellation of the project.

Message

Solvent occlusion is shown to be directly related to the ratio of nucleation rate over crystal growth rate. To minimize the occlusion, fine seed, extended addition time, and better mixing were found to be effective.

EXAMPLE 10-4   *Reactive crystallization of an API*

Goals: To demonstrate the effect of seeding point and addition of reactant on nucleation and filtration rate

Issues: Seeding point, linear versus controlled addition, nucleation, filtration rate

In the reactive crystallization of an API, methane sulfonic acid (MSA) is reacted with the free base of the previous intermediate. Due to the highly compressible nature of the resulting crystals on filtration, caused by fines from the crystallization, it is desirable to minimize the level of secondary nucleation by reducing the degree of supersaturation and allowing a slower release of supersaturation during the reactive crystallization.

The crystallization is conducted at 55°C in a 92/8 v/v ethyl acetate/ethanol system. Table 10-2 outlines the different addition modes of reactant to demonstrate the impact of addition of reactant on the formation of nuclei. Figure 10-12 illustrates the linear and programmed (cubic) addition profiles.

In this example, the impact of seeding point and addition mode of MSA are presented.

Effect of Seeding Point

The influence of the initial MSA/EtOAc charge prior to seeding on the filtration flux is illustrated in Table 10-3. By reducing the percentage of MSA before seeding, the degree of supersaturation is decreased, and therefore the amount of secondary nucleation is reduced. The seed point in the 5%

Table 10-2   Operating Conditions of Original and Modified Procedures

Process parameter	Original procedure	Procedure 1	Procedure 2	Procedure 3	Procedure 4
MSA/EtOAc solution before seeding	35%	5%			
Seed level	2% seed	4% seed			
Remaining MSA/ EtOAc charge (at 55°C)	3 hours linear	5 hours linear	5 hours cubic	8 hours cubic	10 hours linear
Additional age at 55°C	1 hour	0 hour			

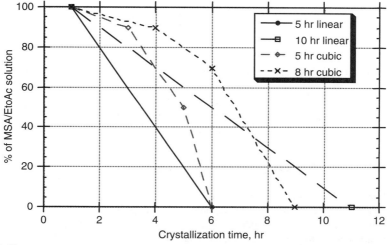

Figure 10-12   Time profiles of different addition procedures.

MSA procedure is very close to the saturation point, which is at ~3 gm/liter MSA salt in 92/8 EtOAc/ EtOH at 55°C.

An in-situ Lasentec FBRM analyzer was used to continuously measure the PSD during the MSA salt crystallization. Figure 10-13 shows the difference in particle population growth between the base procedure (35% MSA) and procedure 1 (5% MSA). While the percentage of total particle counts increased rapidly during the 1 hour seed age period in the original procedure, a corresponding increase was observed in the 5% MSA procedure during the 5 hour linear addition of the MSA/EtOAc solution. This behavior, not evident without the use of an in-situ instrument, suggests that secondary nucleation, rather than crystal growth, still dominates the crystallization. By seeding closer to the saturation point in the 5% procedure, the nucleation is delayed to a region where its rate can be controlled. This suggests

Table 10-3   Effect of Initial MSA Charge on MSA Salt Filtration

Procedure	% MSA before seeding	Supersaturation ratio	Filtration flux (liters/m²/hr)	Increase in filtration flux
Original	35	5.7	383	base flux
1	5	1	943	2.5×

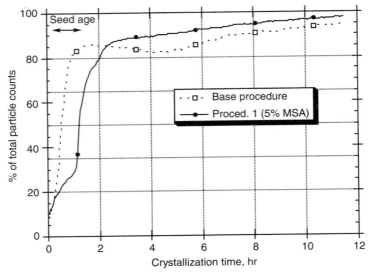

Figure 10-13   Percentage of total particle counts as a function of time—base procedure versus procedure 1.

that the MSA salt crystallization may be further improved by modifying the MSA/EtOAc charge strategy to allow for a slower release of supersaturation.

Effect of MSA/EtOAc Charge Strategy

Figure 10-14 shows the population growth in four crystallization runs. The Lasentec FBRM data reveal that the 8 hour cubic program gave a more gradual slope in the total particle counts curve, indicating the slowest release of supersaturation. The 5 hour cubic addition of MSA/EtOAc gave a population growth profile similar to that of the 10 hour linear addition. Table 10-4 indicates that the 8 hour cubic run indeed gave the fastest filtering slurry.

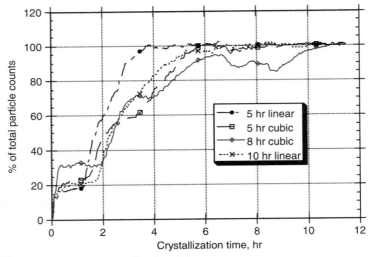

Figure 10-14   Total particle counts as a function of time.

Table 10-4   Effect of MSA Solution Charge Mode on MSA Salt Filtration

MSA/EtOAc charge profile	Increase in filtration flux
5 hour linear	base flux
10 hour linear	$1\times$
5 hour cubic	$1\times$
8 hour cubic	$1.3\times$

Figure 10-14 also points out an interesting characteristic of the reactive crystallization of the MSA salt. As expected, the rate of acid/base reaction between the free base and MSA is much faster than the crystal growth rate. This is illustrated by the sharp increase in the particle counts after the first two step changes in the MSA charge rate ($t = 2$ hours and $t = 4$ hours) in the 8 hour cubic run. Since the MSA/EtOAc solution was charged above surface in these experiments, rapid nucleation may be occurring in local regions of supersaturation. The instantaneous reaction necessitates consideration of local mixing upon scale-up.

Message

This example is another illustration of the important effects of seeding and addition rate on reactive crystallization. In the example, the filtration rate of the product crystals is used as a measure of the impact of these critical variables.

10.5   CREATION OF FINE PARTICLES—IN-LINE REACTIVE CRYSTALLIZATION

In-line methods are applicable to reactive crystallization with systems with a relatively fast reaction rate and a short nucleation induction time. There are two in-line devices that have potential for these applications: impinging jets and vortex mixers. The reader is referred to Chapter 9 for a discussion of impinging jets as applied to antisolvent crystallization. The application to reactive crystallization is similar—creation of high supersaturation in a

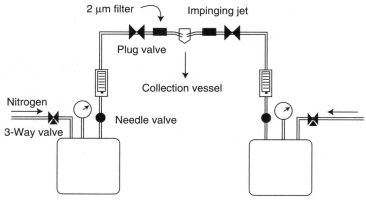

Figure 10-15   Impinging jet crystallization apparatus.
From Condon (2001) with permission.

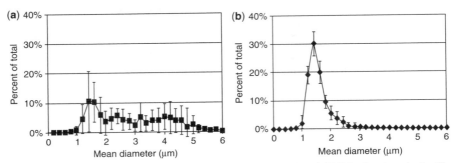

Figure 10-16 Size distribution of calcium oxalate crystallized at (a) 2 and (b) 5.5 m/sec jet velocity. The effect of increased velocity is shown to decrease average particle size and tighten PSD.
From Condon (2001) with permission.

short time. One limitation is in the range of flow rate ratios of the reacting feed streams to balance the momentum of impact. Although some adjustments in feed tube diameter can be helpful in maintaining the momentum balance by equalizing velocities, the ratio of feed streams may be limited to 80 to 20. This limit may be an issue in reactive crystallization applications because the feed streams may have volume ratios of 90/10 or more.

Impinging jets have been studied experimentally for the reactive crystallization of calcium oxalate by Condon (2001) and Hacherl (Condon) (2003). A primary objective of this study was the determination of the capability of the jet system to produce fine

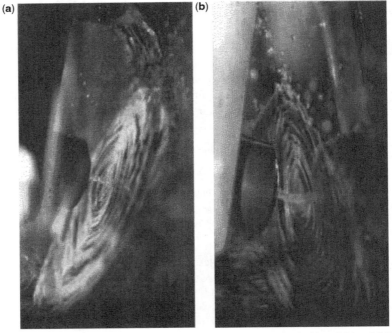

Figure 10-17 Photographs of impingement planes in a free jet (a) at feed velocities of 5.1 and 5.5 m/sec and (b) at 5.5 and 5.1 m/sec of calcium chloride and sodium oxalate, respectively. The intensity of mixing is clearly shown. (See color insert.)

particles. Particles with a mean particle size of 2 μm and a PSD of 1–3 μm were obtained. In addition, the effect of jet velocity on mean particle size and PSD was studied, and a correlation in the expected direction of smaller particle size with increased velocity was obtained. The flowsheet is shown in Fig. 10-15.

This system is complicated by the appearance of two hydrate forms, mono- and dihydrate. The dihydrate is favored at higher excess oxalate local concentrations, while the monohydrate is favored at higher excess calcium local concentrations. Local concentrations within the device were shown to be possible and to cause a shift in hydrate ratio. The particle size results from two typical runs in this study at two jet velocities are shown in Fig. 10-16, and a photograph of the impingement plane is shown in Fig. 10-17 from Condon (2001).

Chapter 11

Special Applications

11.1 INTRODUCTION

While solution crystallization is the main focus of this book, there are applications which are special cases of the crystallization methods described in earlier chapters. For example, a process requiring sterile crystallization is a special example of an antisolvent addition and is presented in Example 11-1. In this particular example, it is necessary to produce fine particles in order to meet drug dissolution time specifications in the final single-dose vials. Furthermore, the material is thermally unstable. Therefore, special precautions have to be taken to minimize any degradation during processing.

Integration of crystallization with other process operations, specifically reaction, is another aspect of crystallization that has significant potential to improve overall process economics. In Examples 11-2 and 11-3, two cases are presented in which products are selectively crystallized during reaction. As a result, reaction selectivity, yield, and the cost of raw materials are significantly improved. Furthermore, solvent recovery is simplified since the same solvent is used in both reaction and crystallization. This type of operation can provide significant improvement in process economics and should be part of the development of crystallization processes when applicable.

Other crystallization techniques that are less frequently applied in the pharmaceutical industry, such as melt and freeze crystallization, may be applicable for some processes. In Example 11-4, purification of dimethyl sulfoxide (DMSO) is presented. In this case, low-level impurities, primarily dimethyl sulfide, are removed by controlled fractional crystallization from the melt (DMSO is a liquid above 18.45°C), in combination with adsorption of impurities from the unfrozen liquid. In the feed DMSO prior to the crystallization step, the impurities, while unacceptable, are at too low a level to be removable by adsorption alone.

In Example 11-5, freeze crystallization of imipenem, which has lower stability in solution at room temperature, is presented. In this process, the product is rapidly frozen to an amorphous solid state to conserve its chemical purity. The temperature is then raised (still below the freezing temperature at this stage), and the amorphous solid converts to a crystalline solid over time. After the completion of the solid-state transition phase, the lyophilization drying cycle is initiated.

Resolution of optical isomers via preferential crystallization is outlined in Chapter 7, Example 7-6, as an example of the use of tightly controlled supersaturation in a cooling crystallization. This process is discussed in greater detail in Example 11-6. The process for resolution of optical isomers utilizes crystallization kinetics, instead of equilibrium solubility, to accomplish the desired isomer separation. It is a proven technique and has been in long-term

Crystallization of Organic Compounds: An Industrial Perspective. By H.-H. Tung, E. L. Paul, M. Midler, and J. A. McCauley

operation at the manufacturing scale. Although this application has a specific focus, the crystallization principles to achieve the required all-growth conditions are broadly applicable.

Other emerging crystallization techniques, such as supercritical fluid crystallization and sonocrystallization (ultrasound in crystallization), are also mentioned. Potential applications for these emerging technologies are present in the pharmaceutical industry.

At the end of this chapter, a brief set of proposed steps is given for the development of a new crystallization process for a pharmaceutical compound.

11.2 CRYSTALLIZATION WITH SUPERCRITICAL FLUIDS

Supercritical fluids (SCFs) are gases and liquids above their critical point. In this state, they are single-phase fluids with some advantageous properties of both liquids and gases. These properties enable them to be used in a unique manner to engineer, or design particles with proper process manipulation.

SCF, while possessing high enough density to solubilize many substances, are highly compressible, especially near the critical point. In many processes using these fluids, solubility of the desired compound can be manipulated by relatively minor changes in temperature and pressure. Additionally, materials dissolved in them benefit by possessing much lower viscosities and higher diffusivities than they do in conventional liquids, allowing very rapid solid phase formation. In fact, this may be so rapid that pains must be taken to avoid the creation of an amorphous, rather than a crystalline, solid (assuming crystallinity is desired).

A comprehensive review of particle design using SCF is provided by Gupta (2006), York et al. (2004), and Jung and Perrut (2001). York (1999) and Subramaniam et al. (1997) give excellent reviews on use of this technology with pharmaceutical compounds.

SCF processing on a large scale had its earliest success with extraction processes. The decaffeination of coffee and tea is carried out in tonnage quantities, and methylene chloride residues thus are eliminated. Extraction of essential oils and flavors has been commercially successful. Because SCF often have only limited solvation properties for some desired materials, SCF processes abound in which nonsupercritical solvents are added at critical points.

Because of its low toxicity and relatively low temperature and pressure critical point, carbon dioxide is the dominant SCF used in the pharmaceutical industry. At its best, an SCF process can produce, in a single step, a pure, dry crystalline solid with high productivity. Much research and development is being carried out in this field.

The common types of SCF processing and their common acronyms, are:

- Rapid expansion of supercritical solutions (RESS): The dissolved product is nucleated by rapid expansion through a nozzle.

- Gas (or SCF) antisolvent (GAS or SAS): SCF is used as antisolvent.

- Aerosol spray extraction system (ASES): Very small droplets are sprayed into SCF antisolvent to produce, micro- or nanoparticles.

- Particles from gas-saturated solutions (or suspensions) (PGSS): SCF is dissolved in liquid product or solution in solvent, followed by rapid depressurization.

One variant of the GAS or SAS process (SCF as antisolvent) is solution enhanced dispersion by supercritical fluids (SEDS). Coaxial nozzles are used to introduce drug solution and carbon dioxide at the desired temperature and pressure. In this case, the SCF carries out both droplet breakup and antisolvent functions. SEDS has been tested for a number of pharmaceutical compounds. As noted above, this is a continuing effort.

11.3 ULTRASOUND IN CRYSTALLIZATION

The benefits of application of ultrasound in crystallization have been reported [by, among others, Ruecroft et al. (2006), McCausland et al. (2001), Thompson and Doraiswamy (2000), Price (1997a), Anderson et al. (1995), Price (1997b), Martin et al. (1993), and Hem (1967)]. Many of the useful effects of sonication are due to cavitation, the opening and subsequent implosion of gas or vapor bubbles. For a further discussion of cavitation and the effects of ultrasound, see Suslick (1988), Suslick et al. (1990), and Young (1989).

Ultrasound can be used beneficially in several key areas of crystallization, such as

- initiation of primary nucleation, narrowing the metastable zone width
- secondary nucleation
- crystal habit and perfection
- crystal size distribution
- reduced agglomeration
- improved product handling

As shown in Examples 7-3, 7-6, and 11-6, ultrasound can be used to break up crystals and alter the crystal morphology. By breaking up crystals along their longest dimension, ultrasound can effectively shorten the aspect ratio of crystals and improve many solid flow properties. In addition, by using ultrasound to initiate nucleation in a controlled manner, crystal growth occurs at a modest supersaturation level. The improvement in product size distribution and reduced agglomeration can potentially result in fewer inclusions of impurities and a number of advantages in mechanical handling.

Despite its potential advantages, several limitations have hindered the wide application of sonication technology beyond the laboratory scale. The first is simply limited understanding of design guidelines for large-scale operation. The transmitted intensity of an ultrasonic probe is confined to a small, poorly defined volume and is not effective over the entire batch volume. The range of ultrasound, even in homogeneous liquids, is limited, and attenuation is further increased by scattering in heterogeneous media, as is the case when crystals are present. Therefore, design parameters for scale-up operations are not clearly defined. At bench scale, installation of in-line sonication horns in recycle lines is a possible aid in evaluating effectiveness on scale-up, although, of course, this may be difficult if the supply of material is extremely limited.

For larger-scale batch processing, further study is required to better understand the relationship among parameters such as sonication power intensity, batch volume, residence time in the recycle loop, and overall time cycle.

Another concern is shedding. The sonication probe tip, normally made of titanium, can suffer from pitting and erosion during use. One commercial solution to this problem is to use a secondary fluid to carry the ultrasonic waves to the product slurry container, as occurs with, for example, glass or metal containers in an ultrasonic bath.

However, as shown in Example 11-6, it is possible to avoid the complication of shedding of horn metal into the product, even with direct contact. This is accomplished by (1) controlling power density at the horn surface (watts/cm^2) and (2) providing proper maintenance of the horns, including periodic inspection of the probe(s) and machining away tip erosion. Erosion on an ultrasonic horn occurs more rapidly on an already damaged surface.

11.4 COMPUTATIONAL FLUID DYNAMICS IN CRYSTALLIZATION

Computational fluid dynamics (CFD) is concerned with obtaining numerical solutions to fluid flow problems by using computers. The equations governing the fluid flow problem are continuity (conservation of mass), Navier-Stokes (conservation of momentum), and energy equations. These form a system of coupled nonlinear partial differential equations (PDEs). Solving a particular problem generally involves first discretizing the physical domain that the flow occurs in, such as the interior of a stirred tank. Commercial CFD software, such as FLUENT, is currently available. It has been used to assess the effect of mixing variables and system geometry on mixing performance. Figure 7-13 shows one example of applying CFD to calculate the flow patterns of two streams in a mixing elbow.

CFD is becoming a more useful tool for describing solid–liquid mixing in crystallization. It can predict the flow patterns, local solids concentration, and local kinetic energy values, taking into account the effects of vessel and agitator shape. Since additional particle population balance equations including crystal growth and nucleation kinetics are required, simplifications such as ignorance of the impact of solid particles on the fluid flow are inevitably applied (Myerson 2002, Chapter 8; Wang and Fox 2004; Woo et al. 2006). Therefore, CFD results need to be examined against the actual experimental data before accepting fully their validity. Despite its limitation, CFD is an excellent educational tool for quick learning and a rapid screening/diagnostic tool for process development.

EXAMPLE 11-1 *Sterile Crystallization of Imipenem*

Goal: Achieve desired physical properties and chemical purity

Issues: Sterile operation, narrow particle size distribution (PSD) small mean particle size (20 microns)

The final step in a multistep synthesis of imipenem, an injectable antibiotic, is a sterile recrystallization.

COMPLICATING FACTORS

In this example, control of growth and narrow PSD are essential because the dried product is filled in single-dose vials and then redissolved for injection at the time of use. The redissolution step in the vial must be rapid (<20 seconds in an aqueous solution) to meet specifications. Therefore, the growth strategy must be designed to achieve a maximum particle size of <20 μm and a narrow PSD to satisfy the rapid dissolution requirement. Milling to reduce particle size is not a desirable option because of temperature instability of the molecule, the need to avoid particulate contamination from the milling operation, and the complexities of sterile milling (delumping the dried product is acceptable because of the lower energy input).

In addition to these specifications, the downstream operations of filtration, washing, and drying require a filterable crystal with low solvent retention in order to facilitate drying at low temperature.

PROCESS DEVELOPMENT

These requirements impose conflicting restraints on the development of a manufacturing process. The common goal of achieving maximum growth for downstream operation efficiency is not consistent with the goal of controlling mean particle size at ~20–30 microns. The most difficult limit to control could be the upper limit of particle size since dissolution of particles or agglomerates >40 microns would be prohibitively slow.

The temperature stability profile of imipenem effectively rules out cooling or evaporative crystallization. Antisolvent addition is the obvious choice. The preferred solvent is water because of the need for sterile filtration of the solution to be fed to the crystallizer. A good antisolvent is known to be acetone. In this case, the antisolvent has the important function of reducing the metastable zone width as well as the solubility.

PREPARATION OF AN AQUEOUS SOLUTION

Issue: Low solubility in water at operating temperatures. The solubility curve is shown in Fig. 11-1, including an estimate of the width of the metastable zone in aqueous solution.

Feed Preparation, Option 1

A solution of 10 gm/liter can be made and concentrated to 30 gm/liter by reverse osmosis (evaporation of an aqueous solution is not viable because of temperature instability).

This solution is supersaturated, as can be seen, but is well within the metastable zone, as shown in Fig. 11-1. This stability under supersaturated conditions is critical for concentration by reverse osmosis and for the following sterile filtration (0.22 micron sterilizing filters).

Following sterile filtration, the antibiotic can be crystallized by addition of acetone.

Option 1 has the serious disadvantage of low yield because of the limit on time and the limit of temperature on concentration by reverse osmosis at the manufacturing scale to prevent unacceptable decomposition. After addition of acetone to the concentrate to crystallize the product, the solubility is unacceptably high for yield purposes.

Feed Preparation, Option 2

A second option is possible only because of the high supersaturation that can be achieved with imipenem in aqueous solution.

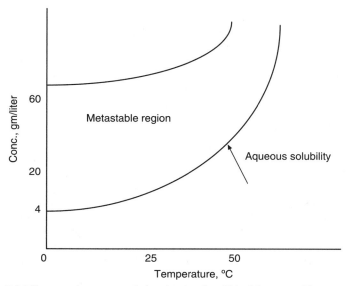

Figure 11-1   Solubility curve in aqueous solution showing the width of the metastable zone.

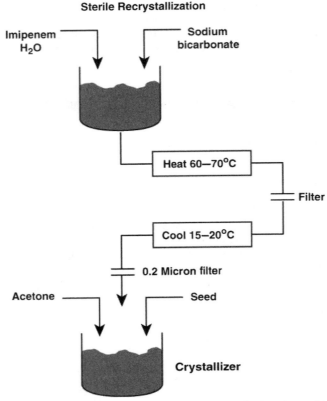

Figure 11-2 Flowsheet for Example 11-1; feed preparation and crystallization of an antibiotic under sterile conditions.

In this option, a higher concentration, 60 gm/liter at 60°C, is achieved by a heat/cool system, as shown in Fig. 11-2. The short contact time for heating to achieve dissolution and cooling to a stable temperature, ~1 minute total, is consistent with chemical stability.

The high supersaturation at filtration temperature, $S = 5$ at 15°C, is acceptable because of the increased width of the metastable zone in aqueous solution combined with a longer induction period—also in aqueous solution. During this time, sterile filtration to the crystallizer without nucleation can be completed—a gift of nature to the development team. Note, however, that premature nucleation from aqueous solution in the feed line to the sterile filter will cause slow sterile filtration and jeopardize successful completion of the run.

In addition, if nucleation from an aqueous solution occurs in the crystallizer—either by excess time before addition of acetone or by subcooling below 15°C in the crystallizer—the resulting crystals will have different crystal morphology. The crystals from an aqueous solution grow rapidly, once nucleated, and produce crystals that dissolve slowly and thereby fail to meet the dissolution time and PSD specifications. An example of the crystals that can nucleate and grow in aqueous solution—before acetone addition—is shown in Fig. 11-3a, where the very large cubic crystals have grown in an aqueous solution, whereas the smaller ones were formed after acetone addition.

Note, however, that these agglomerates show evidence of growing out from a core of an aqueous derived crystal or from drops of oil that are known to form readily before the onset of crystallization.

Note: The initiative of the Merck Plant in Elkton, Virginia, in obtaining Fig. 11-3 and the ones to follow, is gratefully acknowledged. These scanning electron micrographs (SEM) are from recent (2004) production batches of imipenem produced in the sterile area of the plant.

Figure 11-3   (a) The very large cubic crystals have grown in aqueous solution where the smaller ones were formed after acetone addition. (b) SEMs of the final crystals at $100\times$. (c) SEMs of the final crystals at $2000\times$.

The large crystals that appear in the crystallizer cannot be removed and would cause failure of the dissolution time and PSD specifications. Milling and screening cannot be utilized to break or separate the large crystals because of the realistic restrictions of sterile operation.

Crystallization

The imipenem can now be crystallized by the addition of acetone.

Crystallization Option 1: Slow Addition without Seed

All attempts to nucleate without seed failed because of the initial formation of oil followed by uncontrolled formation of amorphous product. This option was pursued because of the complications of sterile powder seeding but without success.

Crystallization Option 2: Addition of Acetone with Seeding and Controlled Growth

Seeding with product from a batch prepared for this purpose was initially utilized, with satisfactory results. However, the PSD was somewhat broader than optimal.

Crystallization Option 3: Use of Milled Seed

As indicated above, milling of this sterile product presents many regulatory and operating problems. An intramuscular dosage form requiring $5-10$ micron particles was subsequently introduced, however, requiring milling. Milled, sterile material with a reproducible PSD then became available for seed.

Final Process

- Acetone, 7% by volume, is added to the sterile-filtered aqueous solution. This amount of acetone dramatically narrows the width of the metastable zone.

- Milled seed ($\sim$0.4%) is added. The ideal amount of seed is calculated to provide the approximate number of 5–10 micron particles that will double in size during essentially all growth (see Chapter 5).

- The remaining acetone is added over a 1 hour period to achieve a final acetone/water ratio of 65/35.

- The crystals are filtered, washed, dried, and delumped, and the bulk product is dry-filled into single-dose vials.

The uniform shape and size achieved by this process are required for these downstream steps to be successfully completed. Key issues are filtration rate, low retention of solvent by the wet filter cake ($<$10%), drying at $<$25°C, and flowability of the delumped crystals.

SEMs of the final crystals at 100$\times$ and 2000$\times$ are shown in Figs. 11-3b and 11-3c.

This process is also successful in achieving growth of crystals that exhibit solid-state stability. Figure 11-4 shows photomicrographs of crystals of this compound that have different stabilities, as indicated. A difference in the sharpness of the end faces is apparent, which is believed to be the cause of this difference. These crystals were grown in isopropanol/water.

Unstable

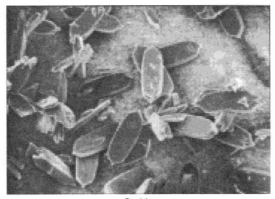

Stable

Figure 11-4   Photomicrographs of antibiotic crystals; the sharpness of the crystal surfaces is an indication of solid-state stability.

EXAMPLE 11-2   *Enhanced Selectivity of a Consecutive-Competitive Reaction by Crystallization of the Desired Product During the Reaction*

BACKGROUND

As introduced in Chapter 10, Section 10.3.8.3, the selectivity of a chemical reaction may be increased above that predicted from its rate constant ratio if the reaction is run under conditions that cause the product to crystallize while the reaction is in progress. A related example is presented in Example 11-3 below. A common system that can be modified by crystallization is the classic consecutive-competitive reaction as follows:

$$A + B \xrightarrow{k_1} R$$

$$R + B \xrightarrow{k_2} S$$

where R is the desired product and S is the over-reaction, undesired product.

For a consecutive reaction that is pseudo-first order in R, the effect of crystallization of R is to reduce the rate of reaction to the undesired over-reaction product S by reducing the concentration of R in the reaction mixture as the reaction proceeds. The topic of increasing reaction selectivity in the presence of, or by the creation of, a second phase, e.g., by crystallization, has been discussed by Sharma (1988) and Paul et al. (2003, chapter 13).

The qualitative effect of crystallizing R during the reaction is shown in Fig. 11-5, where the solution concentrations of the reaction components are shown as a function of mole ratio as B is added to A. In this solvent system, the solubility of R is sufficient to prevent crystallization. Under these conditions, the system is following a homogeneous reaction pathway and the maximum yield of R is determined by the mole ratio and the rate constant ratio, as indicated in equation (11-1) and as shown in Fig. 11-5a.

$$Y_{\exp} = R/A_0 = 1/(1 - \kappa)[(A/A_0)^\kappa - (A/A_0)] \tag{11-1}$$

where $\kappa = k_2/k_1$ and $Y_{\exp}$ is the expected yield under conditions of perfect mixing.

Note: Equation 11-1 applies to perfectly mixed conditions. A decrease in the yield of R (increased S) can result from fast reactions with imperfect mixing, as discussed by Baldyga and Bourne (1999)

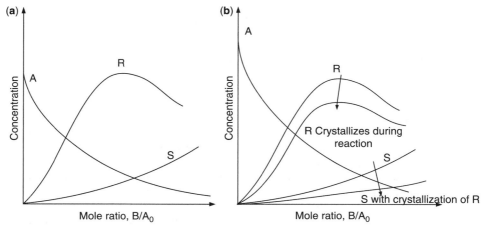

Figure 11-5   Concentration profiles for a consecutive-competitive reaction showing the effect without (a) and with (b) the crystallization of the desired product, R, and the resulting change in overreaction to S.

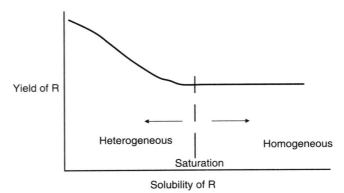

Figure 11-6   Effect of the solubility of the desired product, R, in the reacting system on the yield/selectivity of R.

and Paul et al. (2003, chapter 13). Reactions that are fast enough to be affected by mixing can also be improved by crystallization, but the results may be a function of the mixing issues discussed in these references.

If the solvent system can be changed, or the reaction run at higher concentrations exceeding the solubility of R, R can be crystallized as the reaction proceeds. The selectivity can thus increase, as shown by a decrease in S in Fig. 11-5b. Note that although the amount of R as shown in Fig. 11-5b appears to decrease under these conditions, the R curves are for solution concentration and do not include the amount of R that has crystallized. This effect on yield/selectivity is also shown in Fig. 11-6.

This application of reactive crystallization to achieve increased selectivity is illustrated below.

PROCESS

Goal: Maximize selectivity (yield) of a competitive-consecutive chemical reaction

Issues: The reaction produces a major product, R, which overreacts to an unacceptable amount of by-product, S.

A common method to increase selectivity by limiting the mole ratio of B to A cannot be used because the reaction must be forced to >98% conversion of A because A cannot be removed in subsequent steps.

(*Note*: This restriction in the B/A ratio is not an issue in Example 11-3 below.)

Original Process

One of the reaction steps in a multistep synthesis produces the desired mono-addition product, R, which overreacts to a bis-addition by-product, S in a phase transfer reaction system. The solvent system is water/isopropyl acetate. The reaction is run by feeding reagent B over a several-hour period to a two-phase mixture of the solvents containing the phase transfer catalyst.

As originally run, the amount of solvent was sufficient to solubilize all of the reactants and products throughout the reaction. The rate constants, k_1 and k_2, were determined and the ratio correctly predicted the expected yield using Equation 11-1. This yield was achieved experimentally, and as predicted, the amount of bis (S) formed was unacceptably high ($\sim$18%).

Alternatives

Change the reaction system to reduce k_2, thereby increasing the yield of R and decreasing S. Reacting systems were found that achieved reduced k_2 at a favorable rate constant ratio. However, the primary reaction rate was also reduced, as is common for competitive-consecutive reaction systems, and was too slow for manufacturing purposes.

Reduce the conversion of A, thereby reducing S formation, as shown in Fig. 11-4 at reduced mole ratios. This method is commonly used if downstream processing can remove and possibly recycle unreacted A. However, in this case, downstream separation was not feasible since A could not be separated from R in the subsequent purification by extraction.

Change the apparent rate constant ratio by crystallizing R to reduce its concentration in solution, thereby reducing S formation. This method was investigated, as it offered the best opportunity for success at the minimum change in operations.

Final Process

The solubilities of R and S in the reaction mixture were determined and an appropriate volume of water/isopropyl acetate was found that could solubilize S but allow R to crystallize soon after the start of addition of B to A. The reaction is run in the fed batch (semibatch) mode with B added to A. Two critical crystallization parameters are controlled to ensure the minimum concentration of R (maximum crystallization) during the addition:

- Seed addition as early in the addition of B as possible (as soon as the conversion of A creates an R concentration in the metastable zone).

- Slow addition of B to minimize the concentration of R in solution by allowing time for nucleation and/or growth (particle size is not a key issue in this case because the R is resolubilized after completion of the reaction for the subsequent purification by extraction. R is crystallized only for minimizing S and maximizing yield). The effect of addition rate on yield/selectivity is shown qualitatively in Fig. 11-7, where it can be seen that a critical rate can be determined at which the crystallization rate can maintain the minimum concentration of R. At higher addition rates, selectivity decreases as the R concentration increases.

These process changes resulted in a decrease in S from $\sim 18\%$ to $\sim 3\%$ and a corresponding increase in the yield of R. These results were realized on scale-up to manufacturing.

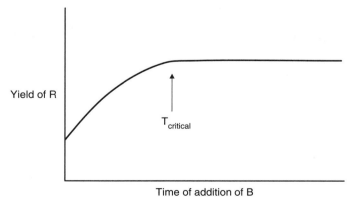

Figure 11-7   Effect of the addition time of reagent B on the yield/selectivity of the desired product, R, resulting from its effect on the crystallization of R.

Message

Crystallization of a reaction product as the reaction proceeds can be effective in increasing reaction selectivity in a complex reaction. It may therefore be advantageous to cause crystallization by changing the concentration or the solvent system, even though there is no other reason to crystallize the product at this point in the synthesis. If crystallization is not possible, addition of an immiscible solvent can also have this "protecting" effect, as discussed in the references above. This process advantage is unique to heterogeneous systems.

EXAMPLE 11-3   *Applying Solubility to Improve Reaction Selectivity*

Goals: To demonstrate the application of solubility and crystallization to improve reaction selectivity and yield

Issues: Solubility, reaction selectivity, yield, crystallization, filtration, recycle

In the synthesis of a drug candidate, one intermediate monoaldehyde (Mono) was formed by reacting two reagents, 7-chloroquinaldine (7-chloro) and isophthalaldehdye (Iso). As shown in Fig. 11-8, reagent Iso has two equally reactive carbonyl groups. One of these can react with the single methyl group of 7-chloro to form the condensed product mono. Since the product Mono has one remaining reactive carbonyl group, it will further react with the starting material Iso to form an inert by-product bis-adduct. This reaction scheme can be modeled by the following consecutive competitive reaction system:

$$A + B \xrightarrow{k_1} R$$

$$R + B \xrightarrow{k_2} S$$

where   A = starting material, isophthalaldehyde (Iso), which contains two reactive carbonyl groups

B = starting material, 7-chloroquinaldine (7-chloro), which contains one methyl group

R = product, monoaldehyde (Mono), which contains one reactive carbonyl group

S = by-product, bis-adduct (Bis), which contains no reacting group and is chemically inert

k_1, k_2 = rate constants. In this reaction system, the ratio of k_1/k_2 is ~ 2 (Tung et al. 1992).

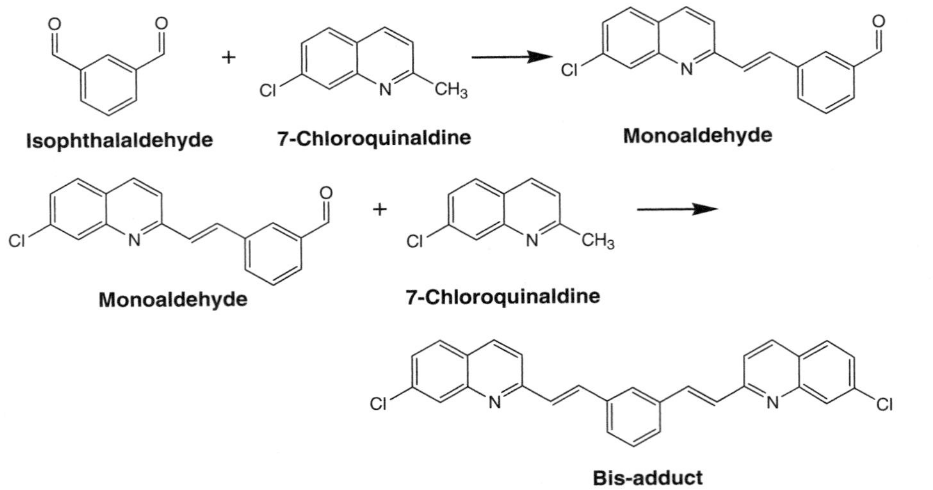

Figure 11-8   Chemical reaction of Example 11-3.

To ensure a high reaction selectivity for this type of reaction, a high A/B ratio is required (Levenspiel 1972).

For convenience and consistency of the presentation, the abbreviated names Iso, 7-chloro, Mono, and Bis will be used in place of the generic A, B, R, and S throughout the rest of this discussion.

Option 1: One-Pass Process

As shown in Fig. 11-9, in the original process, solvent, Iso, and 7-chloro were charged to a reactor first. The batch was heated and aged at the reaction temperature above 100°C. After the reaction was complete, the batch was cooled to approximately 90°C and the majority of by-product Bis was precipitated. The by-product slurry was filtered to remove the precipitated by-product Bis. The filtered clear solution was transferred to a crystallizer and slowly cooled to crystallize the product Mono. The product slurry was filtered, and wet cake was washed and vacuum-dried.

A typical reaction yield of this approach was approximately 73% based upon the charge of 7-chloro. The charge ratio of Iso/7-chloro was 1.5 to 1. Excess Iso was charged to increase the reaction selectivity. However, unreacted Iso was lost in the mother liquor. The final yield after the crystallization was approximate 67%. The isolated cake contained the desired product Mono with approximately 4–5 wt% Bis-adduct. A single solvent (toluene) was used in the process.

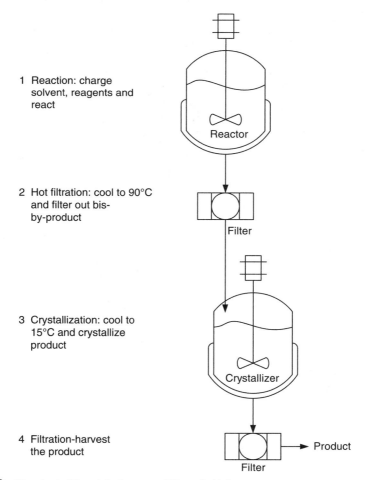

1 Reaction: charge solvent, reagents and react

2 Hot filtration: cool to 90°C and filter out bis-by-product

3 Crystallization: cool to 15°C and crystallize product

4 Filtration-harvest the product

Figure 11-9   Flowsheet of the original process of Example 11-3.

Option 2: Two-Filtration Process with Recycle of Second Filtration Mother Liquor

As mentioned earlier, it is necessary to have a high A/R ratio to ensure high reaction selectivity. This can be accomplished by increasing the charge of Iso over 7-chloro over the current procedure and crystallizing Mono during the reaction. Meanwhile, it is also critical to recover the excess charge of Iso as much as possible in order to minimize the loss of unreacted Iso in the mother liquor. To access the feasibility of alternative options, the first step is to obtain the solubility information on Iso and Mono.

Solubility

Solubility data of the starting material isophthalaldehyde and the product monoaldehyde in one solvent system are presented in Fig. 11-10. It is evident that Iso has a much higher solubility than Mono. The solubility difference between the product and the starting material plays a key role in the design of the two-filtration recycle system. A high ratio of Iso over Mono can be maintained by initially charging a large molar excess of Iso and selectively crystallizing the product Mono during the reaction, while the excess Iso remains solubilized in the reaction mixture. Following the reaction, the recovery of excess Iso is further simplified as a result of the solubility difference. After isolating the cake which contains both the product Mono and the unreacted Iso, the excess Iso can be easily recovered in the filtrate after slurry-washing the cake.

It should be mentioned that the by-product Bis has a lower solubility than both Iso and Mono. Because Bis is chemically inert, its presence does not affect the reaction performance and its solubility data are not included in this communication.

Process Description

Figure 11-11 shows the flowsheet of the two-filtration process with recycle of second filtration mother liquor. For the recycle process, at the beginning of the batch, the starting materials Iso and 7-chloro were charged to a reactor. The Iso charged included the recycled Iso from the previous batch and the fresh Iso. The batch was heated and aged at the desired reaction temperature. After the reaction was complete, the batch was transferred to a crystallizer and cooled over a period of time. Both

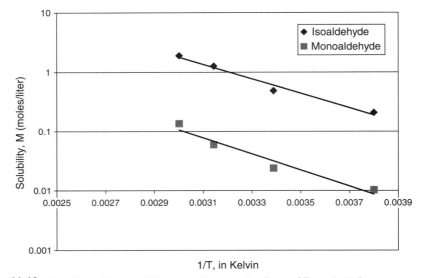

Figure 11-10 Solubility of Mono and Iso in the toluene mother liquor of Example 11-3.

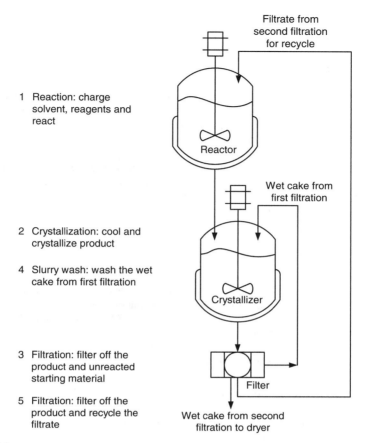

Filtrate from
second filtration
for recycle

1 Reaction: charge
solvent, reagents and
react

Reactor

Wet cake from
first filtration

2 Crystallization: cool and
crystallize product

4 Slurry wash: wash the wet
cake from first filtration

Crystallizer

3 Filtration: filter off the
product and unreacted
starting material

Filter

5 Filtration: filter off the
product and recycle the
filtrate

Wet cake from second
filtration to dryer

Figure 11-11   Flowsheet of the two-filtration process with recycle of second filtration mother liquor.

Mono and the unreacted Iso, along with the by-product Bis, crystallized at this stage. The hot filtration of by-product Bis was omitted due to its reduced formation in the recycle process compared to the original process. The slurry was filtered and the wet cake was washed. The isolated cake contained both Mono and Iso, as well as the by-product Bis.

The isolated wet cake after the first filtration was reslurried in a solvent system which dissolved completely the excess Iso and a small amount of Mono. The slurry was filtered and the wet cake washed. This isolated cake after the second filtration contained Mono and Bis, but not Iso. The mother liquors and washes of the second filtration contained all the unreacted Iso. The second filtration mother liquors and washes were concentrated and recycled to the next batch.

As indicated earlier, the by-product Bis was not removed in the recycle process, primarily due to a much lower level of formation in comparison to the original process. This did not create any additional complication since Bis was chemically inert and could be easily rejected in the next step of the synthesis.

Several key changes were made in the recycle process. First, the level of starting material Iso charged at the beginning of the reaction was increased. Up to three molar equivalents of Iso versus 7-chloro were charged to the batch versus 1.5 molar equivalents charge in the original process. This change significantly increased the reaction selectivity and yield. Second, a slurry wash/filtration operation was included in the recycle process to effectively recover the unreacted Iso. Due to the solubility difference between Iso and Mono, a simple slurry-washing operation recovered all the excess unreacted Iso and reused it in the next cycle. Third, a binary solvent mixture of *n*-heptane and toluene was used which prompted the crystallization of Mono during the reaction and reduced the loss of Mono and Iso in the mother liquor.

Reaction Yield

The experimental results of both the original and recycle processes are listed in Table 11-1. Both the reaction yield and the isolated yield are reported. The reaction yield is equivalent to the isolated yield plus the mother liquor loss in the first filtration.

As shown in Table 11-1, the average reaction yield is only 72.2% for the original process. It is 84.4% for the recycle process.

In the recycle process, the increase in the reaction yield is a direct result of increasing the Iso/7-chloro ratio from 1.5:1 to 3:1, which effectively increases the ratio of Iso/Mono and crystallization of Mono during the reaction. It should be pointed out that due to the difficulty of sampling and filtration of slurry at the reaction temperature, the solid phase cannot be easily assayed to quantify the percentage of Mono present.

Isolated Yield

As shown in Table 11-1, the isolated yield of the recycle process is 82.2% versus 66.1% for the original process. As discussed above, the improvement is primarily due to three factors, all of which contribute to a higher reaction yield: (1) a higher ratio of Iso/Mono, (2) Mono crystallization, and (3) a small amount of Mono in the recycle second filtration mother liquor.

It should be pointed out that reduction of mother liquor loss should be carefully balanced against effective rejection of other minor impurities that are also present.

Raw Material Iso Requirement

As shown in Table 11-1, the fresh charge of raw material Iso for each process is 1.5 and 1.22 equivalents versus 7-chloro, respectively. The excess charge of raw material Iso is reduced from 50% in the original process to 22% in the new process. Since Iso is a key component of the overall process cost, the reduction of Iso requirement provides a substantial saving.

Table 11-1   Experimental Results of the Original Process versus the Two-Filtration Process

Process	Batch no.	Initial mole ratio Iso/7-Chloro/ Mono	Isolated yield, %	Reaction yield, %	Mono purity, wt%	First filtration loss, %	
						Mono	Iso
Original	1	1.5/1/0	64.4	69.6	94.7	5.2	35.3
	2		66.7	73.2	97.6	6.5	44.0
	3		67.4	73.9	95.6	6.5	48.4
Average		1.5/1/0	66.1	72.2	96.0	6.1	42.6
Recycle	169	3/1/0.15	80.6	82.7	89.7	2.1	21.2
	176	3/1/0.09	83.5	85.0	90.0	1.5	20.8
	178	3/1/0.14	81.6	83.9	90.2	2.3	24.3
	180	3/1/0.13	82.9	84.9	88.7	2.0	19.0
	184	3/1/0.13	82.4	84.7	89.7	2.3	20.7
	190	3/1/0.13	82.1	84.4	91.0	2.3	20.2
Average		3/1/0.13	82.2	84.4	89.7	2.1	21.2

Message

By applying solubility information to the reaction, it is possible to improve the reaction selectivity and overall yield. Process integration of crystallization and reaction can be a very powerful tool to improve the process economics.

EXAMPLE 11-4   *Melt Crystallization of Dimethyl Sulfoxide*

Dimethyl sulfoxide (DMSO) is a well-known chemical and has been used as a solvent for many materials in industry. Totally dry DMSO is a liquid above 18.45°C. The excellent solvency of this material led to its use in a variety of pharmaceutical applications. In particular, DMSO was found, in the 1960s, to facilitate the absorption of drugs transcutaneously at a very high rate, and a number of pharmaceutical companies investigated the use of DMSO in this manner.

Unfortunately, commercially available DMSO has a characteristically objectionable odor which is caused by low-level impurities, considerably lower than 1%, largely dimethyl sulfide (DMS) but also a small number of other sulfur-containing compounds. These compounds have an effect on taste as well as smell. They are particularly measurable by UV absorption at 275 mμ. Technical Grade DMSO (pharmaceutically objectionable odor) has a $UV_{275} > 0.25$. Lowering UV_{275} to <0.1 makes DMSO essentially odorless.

Goal: Remove the objectionable impurities

Issues: DMSO absorption of water from the atmosphere, DMSO decomposition when stressed, intense mixing required at the liquid–solid interface

PROCESS OPTIONS

Option 1: Remove the Offending Impurities by Adsorption on a Substrate Such as Activated Carbon or Resin

This option was tested and did not work as a stand-alone process because the DMSO was too dilute in these impurities. The measured adsorption isotherms indicated that unrealistically large amounts of adsorbent would be required, even for marginal results.

Option 2: Remove the Impurities by a Distillation Process

This is the commercial means used by manufacturers of Technical Grade DMSO. The product of these very carefully controlled distillation processes still contains unacceptable levels of DMS and the other odor-causing materials. If continual reworking of the pure DMSO stream is carried out, the impurity concentration stabilizes at a low-level steady-state value, which is still unacceptable by pharmaceutical standards.

The reason distillation is limited in its ability to reach the very low impurities level required is that the highly reactive DMSO decomposes at a finite rate as boilup heat is provided.

SOLUTION

- Freeze out (crystallize) pure solid DMSO, leaving impurities behind in the mother liquors (unfrozen melt).
- Remove the impurities from their higher concentration in the mother liquors by an adsorption process with activated carbon.
- Return the treated mother liquors to the crystallizer.

The above process (Allegretti and Midler 1967) was successful in producing purified DMSO with essentially none of its characteristic odor and $UV_{275} < 0.10$ at a yield of $\sim 100\%$.

Carbon Isotherms

Figure 11-12a presents room temperature adsorption isotherms for some representative activated carbon samples on Technical Grade DMSO, showing the difficulty of reducing UV_{275} to the desired level of <0.10 without additional purification. Figure 11-12a is a schematic of the melt crystallization–carbon column recycle system which was employed to get around this problem. The higher concentration of impurities in the unfrozen melt altered the equilibrium concentration on the activated carbon. In the steady state (Fig. 11-12b) a reasonably sized carbon column could produce effluent suitable for further freeze crystallization, and the yield of the total process was close to 100%.

Freezing Point and Water Removal

DMSO has extreme affinity for water, which was present in feed Technical Grade DMSO at about 0.5% and which was also absorbed from the air despite precautions. For the recycle purification process, water is rejected by the crystalline front of DMSO and accumulates in unfrozen liquid returned to the process. For material balance removal of water, and because the freezing point of DMSO is sensitive to moisture content (Fig. 11-12c), a dehydration step had to be inserted into the recycle loop.

Vapor pressures of DMSO and water are shown in Fig. 11-12d. Based on appearances, they could be separated by a single-stage batch distillation (under vacuum and in a short time to avoid decomposition). Because of the affinity between these two liquids, however, a single-stage water stripper removed too much DMSO in the overhead stream. The solution to this problem was to insert a reflux condenser in the vessel with warm water in the jacket to allow return of the entrained DMSO in the vapor stream back to the boilup. In accord with Fig. 11-12d, the dehydration system pressure was maintained at about 50 mm Hg and the condenser jacket temperature was set at about 50–60°C, which successfully removed the water and maintained the freezing point near 18°C in the crystallizer.

Mixing in the Freeze Crystallizer

Previous discussion in this book, primarily in Chapter 6, points out the necessity for controlling conditions in the localized region where nucleation and crystal growth processes are taking place. The region of concern in the growing DMSO crystallizer was at the advancing front of the freezing product. In order to ensure effective heat and mass transfer at the growing front, a crystallizer was fabricated (Fig. 11-12e) with an outer jacket and a center tube containing flowing coolant. The temperature driving force for crystallization was held at about 5°C (13°C cooling surface, 18°C freezing point). A double vertical impeller moved the moderate-viscosity fluid rapidly for the entire length of the center tube and jacket to ensure effective rejection of impurities.

Approximately 50% of the frozen DMSO was crystallized in each cycle. After removal of the unfrozen impurities concentrate, the frozen (odorless, low-UV_{275}) product was thawed and packaged.

Early in development, a roller drum freeze crystallizer was constructed in which the melt container was fitted with an ultrasonic transducer to sonicate the growing DMSO ice front. Although the unit worked mechanically, the ultrasonic energy caused excessive DMSO decomposition, and the more conventional stirred crystallizer described above was fabricated.

Message

Crystallization from the melt, often considered in recent decades for avoidance of solvents, can be effective in manufacturing very high purity product even when solvents are not an issue. The need to pay attention to mass and heat transfer principles, and to provide suitable agitation to meet system requirements, bears many similarities to more conventional crystallization from solution.

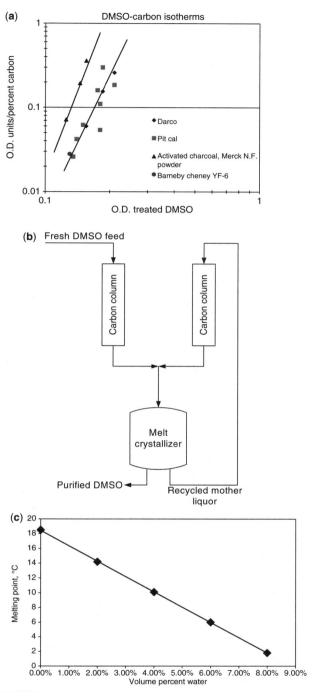

Figure 11-12   (a) DMSO and active carbon absorption isotherms at room temperature. (b) Schematic of the melt crystallization with a carbon column recycle system. (c) Effect of water on the freezing point of DMSO. (d) Vapor pressure of DMSO and water. (e) Crystallizer for melt crystallization of DMSO.

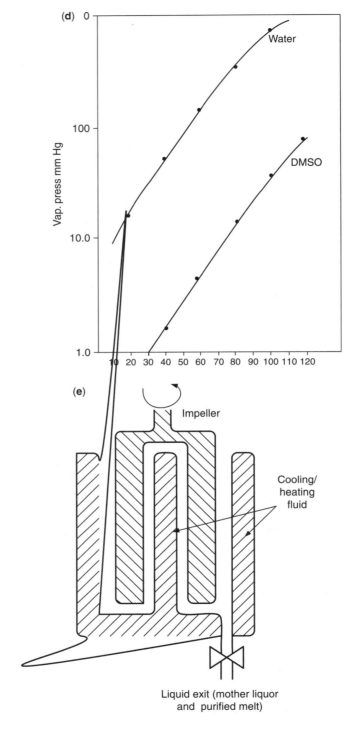

(d)

Vap. press mm Hg

Water

DMSO

(e)

Impeller

Cooling/
heating
fluid

Liquid exit (mother liquor
and purified melt)

Figure 11-12   *Continued.*

EXAMPLE 11-5 *Freeze Crystallization of Imipenem*

Goals: To investigate the formation of crystalline imipenem from amorphous imipenem during freeze crystallization

Issues: freeze drying, freeze crystallization, solid–liquid (S–L) equilibrium, amorphism, and crystallinity

Freeze drying, or lyophilization, is useful in the pharmaceutical industry because compounds which are heat-sensitive and exhibit poor stability in solution can, in many cases, be freeze-dried in order to prepare rapidly soluble and sterile injectable pharmaceuticals with minimal degradation. Basic studies of freeze drying include the work of Deluca and Lachman (1965) on the physical-chemical parameters of eutectic temperature and solubility, MacKenzie (1965) on sublimation, MacKenzie (1977) on the physical-chemical basis for freeze drying, MacKenzie (1977) on nonequilibrium freezing behavior of aqueous systems, MacKenzie (1985) on the fundamentals and applications of freeze drying, and Franks (1990) on predicting successful freeze drying.

Freeze-dried samples, however, are often found to be amorphous, exhibiting a low degree of crystallinity as determined by X-ray powder diffraction analysis. For some pharmaceuticals, a low degree of crystallinity decreases stability, as in the case with imipenem, which, although readily soluble in this amorphous state, is unstable and readily decomposes. Thus, it is of great interest to study and modify freeze-drying procedures in order to induce crystallization during processing of pharmaceutical entities (Cise and Roy 1979).

Option 1: Direct Lyophilization

In the process for the preparation of imipenem (Fig. 11-13), the drug is first dissolved in a solvent mixture of acetone/water, sterile-filtered into single-dose vials, and frozen at a temperature of $-40°C$. Sodium bicarbonate is included in the solution make-up to improve the chemical stability of imipenem. Freezing is observed to take place within 15 minutes of the solution's being charged to the freeze dryer, resulting in the formation of a solid matrix of frozen solvent and amorphous drug. The lyophilization cycle is initiated by placing the system under vacuum, and the temperature is slowly raised to remove all the frozen solvent by sublimation. Imipenem solid derived from this process is 100% amorphous by X-ray powder diffraction analysis (Crocker and McCauley 1995).

Option 2: Freeze Crystallization

This process is described by Connolly et al. (1996).

Similar to the process described in Option 1 above, the antibiotic is first dissolved in a solvent mixture of acetone/water and frozen at a temperature of $-40°C$. The freeze crystallization cycle is then started by raising the temperature to between $-10°C$ and $-30°C$ and holding the system for

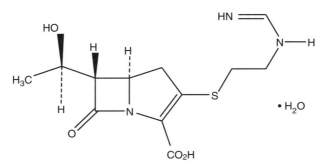

Figure 11-13 Chemical structure of imipenem.

Table 11-2   Freeze Crystallization Cycles (All Acetone Levels)

Stage	Starting temperature, °C	Ramping time, hr	End temperature, °C	Aging time, hr
1	−40°C	1	−10, −20, −30	10
2		5	−30	0
3		1	−40	99.95

several hours. The system is then frozen again to below −40°C. The lyophilization cycle is initiated by placing the system under vacuum, and the temperature is slowly raised to remove all the frozen solvent by sublimation.

Table 11-2 shows the freeze crystallization cycles for the study. Three different freeze crystallization temperatures, −10°C, −20°C, and −30°C, and three different levels of acetone, 10 vol%, 20 vol%, and 30 vol%, were studied. The output variables were the degree of crystallization of imipenem products and the percentage of the liquid phase at the freeze crystallization temperature. For the aging time of stage 3, a value of 99.95 hours was shown in order to hold the freeze dryer shelf temperature at −40°C overnight.

DSC Thermograph

Figures 11-14 to 11-17 show the differential scanning calorimetry (DSC) thermograph of the acetone/water solvent mixture with and without the presence of imipenem. The DSC curves indicate endothermic peaks which correspond to the phase transitions. The peaks around 0°C and −95°C represent the melting of the solid water and acetone phases, respectively. The peaks around −19°C represent the melting of the solid clathrate phase (Rosso et al. 1975). The clathrate is a solid phase complex of 17 water molecules surrounding a single molecule of acetone, $17H_2O^*(CH_3)_2CO$. In addition, the DSC thermograms indicate only slight temperature fluctuations for the endotherms with the addition of imipenem and the sodium bicarbonate. Therefore, the acetone/water binary phase behavior is not affected significantly by the addition of imipenem and sodium bicarbonate.

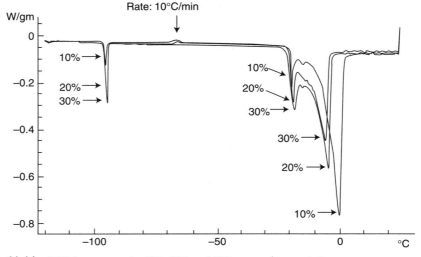

Figure 11-14   DSC thermograms for 10%, 20%, and 30% acetone/water solutions.

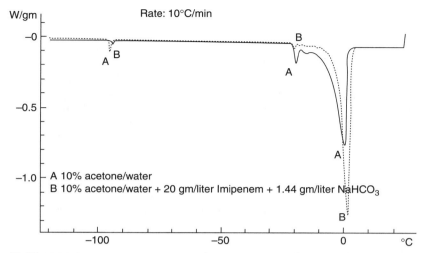

Figure 11-15   DSC thermograms for 10% acetone/water and imipenem/NaHCO$_3$ in a 10% acetone/water solution.

Crystallinity of Imipenem and Phase Behavior of Acetone/Water

The results of the crystallization trials are summarized in Table 11-3. The acetone/water binary phase diagram was used to approximate equilibrium conditions for the solvent systems at the freeze crystallization conditions in the absence of a multicomponent phase diagram for the acetone/water/imipenem/NaHCO$_3$ system.

These results indicate some obvious patterns of crystallization. High degrees of crystallization occur independently of the initial acetone composition of the solvent at −10°C, and the observed degree of crystallinity decreases significantly as the freeze crystallization temperature is lowered. In addition, at the lower crystallization temperature, the observed degree of crystallinity increases with increasing initial acetone composition of the solvent. This trend is particularly significant at −30°C.

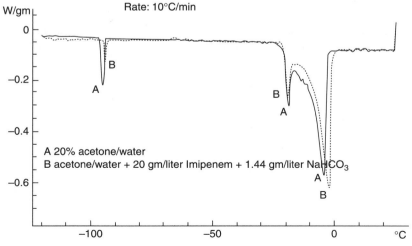

Figure 11-16   DSC thermograms for 20% acetone/water and imipenem/NaHCO$_3$ in a 20% acetone/water solution.

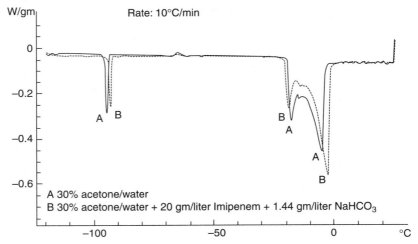

Figure 11-17   DSC thermograms for 30% acetone/water and imipenem/NaHCO$_3$ in a 30% acetone/water solution.

The degree of crystallinity increases from 21% to 80% as the initial acetone composition of the solvent is increased from 20 vol% to 30 vol%.

Using the acetone/water binary phase diagram (Rosso et al. 1975), the relative weight percentages of liquid and solid phases observed at equilibrium were calculated as shown in the last column of Table 11-3. According to the DSC curves and the phase diagram, the formation of a solid clathrate occurs at $-19°C$. Since two freeze crystallization temperatures at $-30°C$ and $-40°C$ are well below the clathrate formation temperature, the weight percentages of the liquid phase observed at these two temperatures were reported, assuming that the clathrate forms. However, for the crystallization temperature at $-20°C$, clathrate formation is assumed not to occur since the temperature of the system fluctuates under the experimental conditions.

The calculated weight percentages of the liquid phases observed at equilibrium provide further insight into the observed crystallization patterns. As shown in Fig. 11-18, the data suggest a direct correlation between the presence of a liquid phase during the freeze crystallization stage and the degree of

Table 11-3   Imipenem Crystallinity and Phase Behavior of Acetone/Water

Acetone, vol%	Freeze crystallization temperature, °C	Degree of crystallinity, %	Liquid level wt% in acetone/water system
10	-10	99	32.2
	-20	78	12.6
	-30	8	0
	-40	6	0
20	-10	99	64.4
	-20	77	26.3
	-30	21	0.1
	-40	14	0.1
30	-10	100	96.6
	-20	93	39.4
	-30	80	14.4
	-40	25	12.8

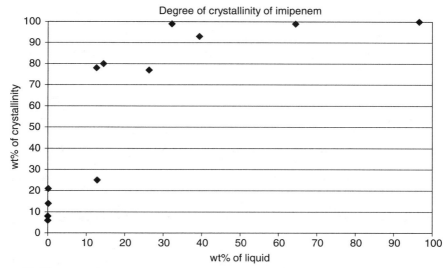

Figure 11-18   Relationship between the degree of crystallinity of imipenem and the equilibrium wt% of liquid in the system.

crystallinity of the resulting product. For imipenem crystallized under conditions in which significant levels of liquid are present at equilibrium, high degrees of crystallinity are observed, and vice versa.

The solubility values for imipenem in the 30% acetone/water solution are 0.72, 0.40, and 0.22 gm/liter at $-10°C$, $-20°C$, and $-30°C$, respectively. Values for solubilities in the equilibrium solutions at $-20°C$ and $-30°C$ could not be measured. These solubility results indicate that imipenem is slightly soluble in the liquid phases present at equilibrium under the freeze crystallization conditions. Therefore, it may be assumed that the transition from amorphous form to crystalline form is facilitated by the presence of a liquid phase.

Message

Transition of amorphous imipenem to crystalline material can be greatly facilitated in the presence of liquid even at the frozen stage. The degree of crystallinity shows a clear correlation with the degree of the liquid level in the frozen solution.

EXAMPLE 11-6   *Continuous Separation of Stereoisomers*

Goals: Kinetic separation of isomers by crystal growth of the desired isomer on the seed bed of that isomer (desired: all-growth process)

A robust, continuous process applicable to selective crystallization in general

Issues: Minimize nucleation of the undesired isomer—strict control of supersaturation

Maintain particle balance—create new growth centers in the absence of nucleation

As noted in the introductory section of this chapter, this process is described in Chapter 7, Example 7-6. It is included in this chapter in greater detail for those who wish to utilize it in a development project.

Many pharmaceutical products are specific stereoisomers, and in many cases the inactive enantiomer must be removed to minimize side effects. In recent years, advances in selective stereochemistry have reduced the need for isomer separation. However, many synthetic processes continue to have racemic output (desired and undesired stereoisomers).

Separation of these products reduces the possibility of side effects from the inactive isomer. In addition to the reduction of side effects, separation of these isomers often provides the possibility of greatly improved yield if the inactive entity can be racemized and recycled.

Because enantiomers of the same compound have many of the same thermodynamic properties, including solubility characteristics in nonchiral solvents, they are particularly difficult to separate in equilibrium processes.

OPTIONS

Option 1: Chiral Chemistry to Produce Only the Desired Isomer

This is a desirable solution, but it is not always feasible or economically viable.

Option 2: Use of a Chiral Additive to Create a Diastereoisomeric Set of Compounds

Diastereoisomers have two chiral centers, and those created with the compound being separated will not have the same solubilities and other properties as the original isomers. This is the most common option exercised in practice. However, it can often be quite costly, because the chiral compound being added to carry out the separation is frequently very expensive, and even in a recycle operation, it can add much cost due to material loss.

Option 3: Use of Membrane, Enzyme, or Other Separation Technology

This is also desirable but often costly.

Option 4: (Solution)

- Kinetic separation of enantiomers in heavily seeded, all-growth, limited residence time crystallizers.
- A continuous (steady-state) process for tight control of supersaturation, unchanging with time.
- Fluidized bed crystallizers to avoid the need for heavy-magma, high-flux filtration equipment.

Kinetic resolution is the least expensive alternative, compared to chiral chemistry, creation of diastereoisomers, or separation devices such as membranes. Additionally, it produces the highest optical purity product of all these methods.

In general, kinetic resolution requires the presence of a racemic mixture (conglomerate) and the absence of a (generally lower-solubility) racemic compound (both enantiomers in the crystal lattice). This is not always the case, however, and depending on the relative rates of nucleation and crystal growth of the respective forms, a kinetic (nonthermodynamic) isomer separation can sometimes be effected even when a racemic compound is possible. In the case of solid solution of the enantiomers (no lattice fit requirement), an equilibrium process will essentially always be required.

Continuous kinetic resolution of stereoisomers was put into production, using stirred tank crystallizers, at Merck & Co. in 1962 magazine-[*Chemical Engineering* staff (1965); Krieger et al. (1968); Lago et al. (1966)]. New chemistry was accompanied by a switch to production-scale fluidized bed crystallizers in 1967, using ultrasonic cleavage of seed particles to maintain population balance [Allegretti and Midler (1970), Midler (1970, 1975, 1976)]. High-speed film studies on crystal cleavage in the ultrasonic field were described by Klink et al. (1971). A patent by the Ajinomoto Company [Ito et al. (1965)] using column crystallizers to resolve amino acids indicated that visible nucleation in the crystallizers was acceptable, which is contrary to the experience at Merck.

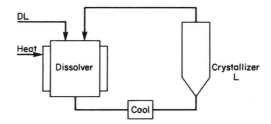

(a) Single crystallizer, semibatch operation

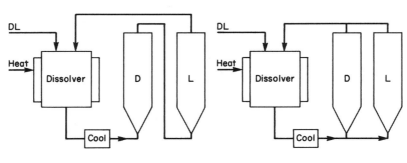

(b) Two crystallizers in series   (c) Two crystallizers in parallel

Figure 11-19   Flowsheet of stereoisomer resolution systems.

Stirred tank and fluidized processes for stereoisomer resolution share the following characteristics:

- All locations in the process stream (including the walls of heat exchangers) are maintained at low supersaturation, well within the metastable region, to preserve isomeric purity within each seed bed by avoiding nucleation.

- All configurations contain a provision for generation of new seed particles to replace those removed in the product harvest streams. Generally, this is carried out by natural attrition, as, for example, in the high shear regions within recirculation pumps, or with superimposed attrition by devices inserted for this purpose.

Typical flow diagrams for both stirred tank and fluidized bed resolution systems are shown in Figs. 11-19 to 11-21.

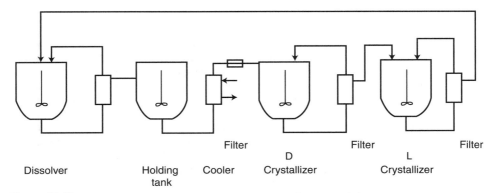

Figure 11-20   Configuration of a stirred tank (series flow) stereoisomer resolution system.

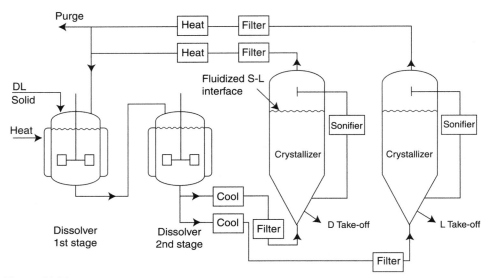

Figure 11-21   Configuration of a fluidized bed (parallel flow) stereoisomer resolution system.

Figure 11-19 shows three flow systems capable of carrying out stereoisomer (or other) resolution by crystallization:

- The *single-crystallizer* system has the benefit of simplicity but requires periodic harvesting of the dissolver.

- *Series operation* always has the desired product in the downstream crystallizer. It decreases the chance for isomer contamination in that unit by reducing supersaturation of the unwanted isomer.

- *Parallel operation* is easier to run because both crystallizers are operated the same way. Harvesting and other operations carried out in one operation do not affect the other, and any washout of crystals from either crystallizer returns to the dissolver rather than possibly contaminating the other isomer seed bed. For these reasons, parallel operation is generally the method of choice.

Figures 11-20 and 11-21 show configurations for stirred tank (series flow) and fluidized bed (parallel flow) systems which have been run at factory scale.

Fluidization

Design of fluidized bed crystallizers requires estimates of the required seed bed volume and the quality of fluidization. Fluidization behavior of seed was measured in the laboratory for monosized cuts. For a given volumetric flow rate through any given fluidized bed crystallizer, there is a minimum and maximum particle size which will result. The minimum is that below which the particles will elutriate out the top of the column. The maximum is determined by the size at which removal or controlled attrition (see below) takes place at the bottom.

It is desirable to measure the fluidization behavior of the particular solids in the column, ideally in actual mother liquors. A typical set of curves is shown in Fig. 11-22a. Each particle size is a relatively monosized sieve fraction. The x-intercept of each curve is the terminal settling velocity (slurry concentration 0 gm/liter), but it is more accurate to measure the terminal settling velocity separately with a stopwatch and a graduated cylinder. Such a data set is shown in Fig. 11-22b.

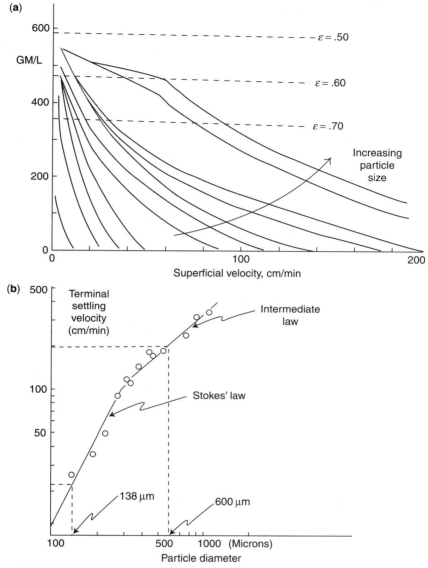

Figure 11-22   (a) Fluidization behavior of particular solids in a (liquid) fluidized bed. (b) Terminal settling velocity of particles in liquid.

The solid in liquid-fluidized bed systems, as distinct from most fluidized with gas, is generally suspended uniformly. Liquid flow, with reasonable column geometry, can be uniformly mixed radially with minimal backmixing. Even in a column of constant diameter, the particles tend to classify according to size (largest at the bottom). Plug flow can be further enhanced by having a substantial tapered section in the bottom.

The crystallizers described in this section were designed with a 3:1 diameter ratio (9:1 ratio in linear velocity). Plug flow approximation for these fluidized bed crystallizers was confirmed by point insertion studies with a radioactive tracer.

Crystal Growth

The fluidization studies noted above for monosized crystal populations showed that the minimum particle size in the fluidized bed crystallizers should be 150 microns and the maximum 600 microns. In an all-growth process (absence of nucleation), the smallest should grow to become the largest with a distribution like that of the "Idealized" curve in Fig. 11-23. The histogram plot of actual column data (Actual) in the same figure shows that the fluidized bed crystallizer run under the conditions of that experiment, corresponds to an all-growth situation.

Growth rate kinetics for most of the stereoisomer resolution processes run at Merck have been essentially first-order with respect to supersaturation. Figure 11-24 shows an exception (2.5 order). The parallel plots in Fig. 11-24 are for different acid concentrations used to solubilize the compound, while supersaturation was generated by a temperature difference between the dissolver and crystallizer.

For one of the stereoisomer separation systems, the (first-order) growth rate constant was measured for both the stirred tank (CSTR) and fluidized bed flow systems, with and without a nucleation suppressant. The data are shown in Table 11-4.

In this case at least, the fluidized bed system showed considerably faster growth rate kinetics than the stirred tank, presumably by convective enhancement in the boundary layer on the surface of the crystals. Additionally, since nucleation suppressants nearly always also inhibit the growth rate, the slower growth kinetics in this case are not unexpected.

Population Balance Control

The astute observer will see that the fluidized bed crystallization system shown in Fig. 11-21 has an unusual feature. Flow sonication units are in position to operate on pumped slurry from the seed beds. In other systems, the sonicators are located internally in the bottom of the column. The "sonifiers" are power ultrasonic horns, which create sufficient cavitation and impact energy to break the crystals along cleavage planes.

Sonication is the means used in the fluidized bed crystallizers for maintaining the number of seed particles in the magma to replace those removed in the product and concurrently preventing the

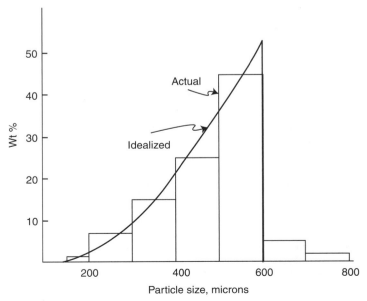

Figure 11-23   The histogram plots the actual column data versus the Idealized curve.

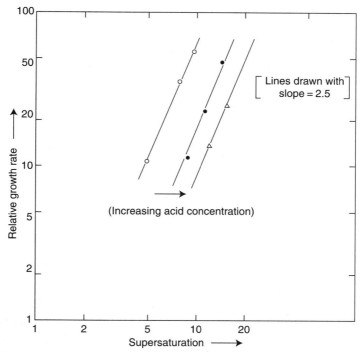

Figure 11-24   An exception (2.5-order) of relative growth rate as a function of supersaturation for different acid concentrations.

formation of overly large crystals. Excessive particle size starves the seed bed of crystal surface area for growth and, in the case of fluidized bed crystallizers, causes sluggish solids movement, which can cause the particles to grow together.

Since separation of stereisomers requires that this type of crystallization be all-growth, some means for creation of new seed particles must be provided for any crystallizer type. These particles can be added from an external source or can be created internally by attrition. An internal attrition process would be preferred to minimize the number of operations.

Stirred tanks with a recycle loop have natural attrition from both the pump and the agitator. The extent of this breakdown is clearly dependent on the types of pump and impeller employed. Fluidized

Table 11-4   First-Order Growth Rate Constants (Fluidized Bed versus CSTR Crystallizer System)

| | $k \times 1000$: gm/hr, cm^2, gm/cm^3 | | |
| | Fluidized bed | CSTR | |
Crystallizer temp. (°C)	No nucleation suppressant		Nucleation suppressant added
25	40		2.2
35	70	16	2.3
41	90		2.3
46			4.1

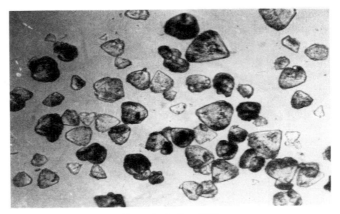

Unsonicated crystals

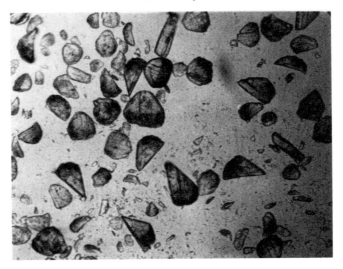

Sonicated crystals

Figure 11-25 The unsonicated crystals are disrupted by sonication along cleavage planes.

Figure 11-26 Sonicators located at the bottom of the column.

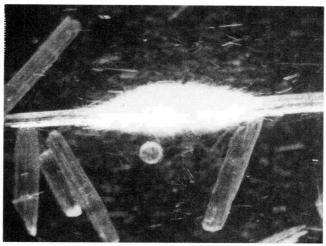

Figure 11-27   Fines generated by sonication can adhere to other particles.

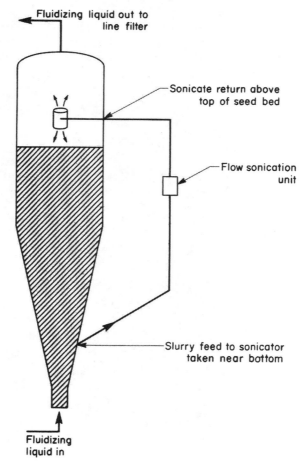

Figure 11-28   The sonicator is an external circulation loop.

beds have a special requirement: any attrition process must minimize fines, because they wash out of the seed bed by elutriation and thus reduce the surface area for growth. A study of possible attrition and wet-milling options showed that sonication created the fewest fines because of the nature of crystal disruption.

Figure 11-25 shows the nature of that disruption. The unsonicated crystals (top) are disrupted by sonication along cleavage planes (bottom), a mechanism which produces few fines to elutriate out of the seed bed. For this product, the sonicators were located in the column bottom (Fig. 11-26). Whether the sonicators are located internally, as shown, or in a slurry flow system, as shown in Fig. 11-28, the crystal population fed to the sonicators is always drawn from the bottom of the seed bed, where the largest particles are located.

Some crystals are more friable than those shown in Fig. 11-25, particularly when they grow in a relatively needle-shaped morphology. In this case, more fines may be produced. In addition to creating some washout from the bed, the fines can adhere to the other particles, as shown in Fig. 11-27. The fines can completely coat the particles, and the resulting seed bed becomes exceedingly dilute and unproductive. This issue was successfully dealt with by installing the external flow sonication system, with the slurry return above the top of the fluidized seed bed, as shown in Fig. 11-28. In this system, whatever fines are produced wash out at the top rather than passing through the bed,

Figure 11-29   Seed crystals entering (top) and leaving (bottom) the flow sonicator.

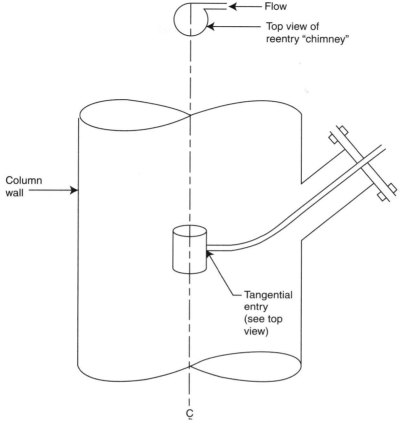

Figure 11-30   Slurry return fines disengagement can be enhanced by a tangential flow device.

and the coating of particles is minimized. Depending on the actual amount of fines produced, they can either be filtered as product (they are pure isomer) or carried back to the dissolver for resolubilization.

Figure 11-29 shows seed crystals entering (top) and leaving (bottom) the flow sonicator. Although not necessary, slurry return fines disengagement can be enhanced by a tangential flow device like that shown in Fig. 11-30.

Fluidized Bed Crystallizer Scale-up

When scaling up column geometry, it is desirable to know the localized slurry concentrations within the crystallizer. Excessively high concentrations can result in sluggish fluidization and possible agglomeration, and excessively low concentrations decrease productivity. These local concentrations can be calculated, assuming a classified bed (largest particles on the bottom), and either a measured or estimated size distribution.

The results of these calculations for the particles shown in Fig. 11-25 are shown in Fig. 11-31. For the laboratory (3 inch diameter) column shown at the top, the maximum slurry concentration (most sluggish fluidization) was calculated to be in the region just above the top of the tapered section. This corresponded to visual observation of operation in the (glass) columns. Pilot plant crystallizers were designed with a similar geometry.

Comparable calculation of a factory column with a similar ratio of taper volume to that in the straight section indicated that fluidization would be excessively sluggish in parts of the crystallizer,

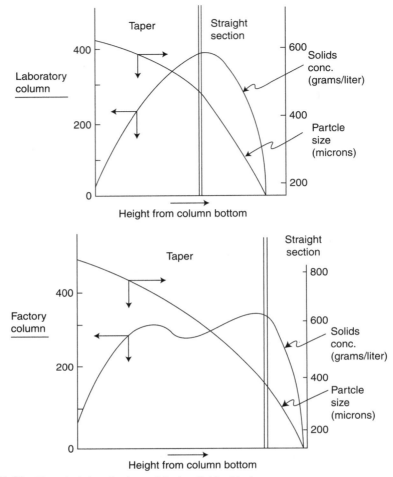

Figure 11-31   The calculations for the particles in a fluidized bed.

and the relative taper volume was increased. Calculation of the ultimate plant design localized slurry concentrations is presented at the bottom of Fig. 11-31, showing a relatively uniform slurry concentration. Samples taken from various taps in the factory operation confirmed these results.

    To estimate the amount of ultrasonic energy required to run in the scale-up crystallizer, a study was carried out measuring crystal disruption by elutriation from a glass flow unit. As shown in Table 11-5, breakage was essentially first-order with sonication power and slurry concentration.

Table 11-5   First-Order Rate Constants for Ultrasonic
Crystal Breakage $\dfrac{1}{P}\dfrac{dm}{dt} = k'$ (slurry-concentration)

P (watts)	k' [gm/hr per (gm/liter, watt)]
30	0.031
55	0.028
68	0.029

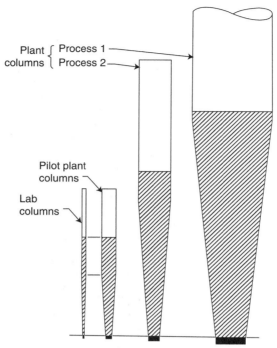

Figure 11-32   Different relative scales of some actual fluidized bed crystallization operations.

Figure 11-33   A sight glass view of a sharp interface in a factory-scale crystallizer.

This also implies that the main mechanism for crystal breakage is individual crystal breakage in the cavitation field rather than particle-particle interaction, which is more common for most wet-milling operations. The "breakage constant" of about 0.03 gm/hr per (gm/liter, watt) was found to hold for a number of small molecule organic compounds tested in the fluidized bed crystallizer system.

As noted above in the section on fluidization, it is appropriate to minimize the angle of the conical section from the vertical to minimize backmixing. Different relative scales of some actual fluidized bed crystallization operations are shown in Fig. 11-32.

In a growth-dominated fluidized bed crystallizer, the interface between the fluidized bed and the mother liquors should be sharp, as the particles either grow larger or wash out at the top. Figure 11-33 shows a sight glass view of such an interface in a factory-scale crystallizer.

As presented earlier in Fig. 7-30, it shows a pair of factory fluidized bed crystallizers, one for each stereoisomer, in construction. These correspond to the calculation at the bottom of Fig. 11-31. Internal sonicators (Fig. 11-26) were installed in the column bottoms. The blowers shown at the bottom of each column were used to cool the sonication units.

Message

When close control of an operating parameter (e.g., supersaturation) is required for a process, a continuous operation should be considered. The success of sonication in this case proves, once again, that the manufacture of high-value products (such as pharmaceuticals) can often tolerate technologies which are unusual in the production of less expensive materials.

11.5 STRATEGIC CONSIDERATIONS FOR DEVELOPMENT OF A NEW CRYSTALLIZATION PROCESS

Below is a brief set of guidelines for factors to consider, and proposed activity, when developing a new crystallization process for a pharmaceutical compound.

Development step	Developmental action
Laboratory observations Procedure makes an oil, gum, or amorphous solid	• Purify a sample by the most applicable technique, e.g., Prep HPLC. • Concentrate the purified sample to dryness. • Choose two or three solvents and possible mixtures with solubilities >20 gm/liter if possible. • Dissolve the residue by heating as appropriate. • Cool, concentrate, and/or add antisolvent to create low supersaturation. • Induce nucleation by seeding with foreign particles.

Note: These suggestions are only a first step in what may be a long series of trials to induce initial crystallization. The most important step may be removal of impurities that may inhibit nucleation.

Crystalline solids are prepared	• Analytical evaluation for crystallinity, hydrates, solvates, as appropriate. • Prepare crystalline material to be used as seed in the experimental program. • Determine the solvent or solvent mixture candidates.

Note: If this is the step for preparation of an API, the solvent(s) selection must include consideration of FDA guidelines

Solvent screening	• Determine the solubility in the chosen solvent and solvent mixtures as a function of temperature and composition in pure solvent(s) and in process mixtures.
	• Determine the impurity profile and evaluate rejection by crystallization.
	• Determine if any impurities form solid solutions with the product.
	• Evaluate the most probable appropriate crystallization method based on solubility, crystallization kinetics, and metastable zone width, if needed.
	• Evaluate analytical characteristics including thermogravimetric analysis (TGA), DSC, etc., as appropriate and prepare a sample for unit cell evaluation and other analytical information as required.
	• Determine target mean particle size, PSD, etc., as required for subsequent step.

Engineering laboratory development	Developmental action
Set objectives for physical attributes and chemical purity	• For isolation of an intermediate, design crystallization to maximize growth and for facile integration with following step.
	• For an API, a target mean particle size, PSD, and purity will be set by formulation specifications.
	• Determine feasibility to achieve the required physical attributes by vessel crystallization and milling as needed.
Experimental apparatus	• For vessel crystallization (mean particle size >10 microns): Prepare experimental crystallizer and accessories for minimum >1 liter experiments with the ability to change impellers, speed, and position of antisolvent addition, and with provision for baffles and appropriate online measurement tools, including FBRM and FTIR.
	• For other methods, see Section 9.2 on in-line crystallization (mean particle size <10 microns) and Section 4.3.2 for a discussion of semicontinuous fluidized bed growth rate measurement apparatus.
Experimental program	• Duplicate laboratory procedure and evaluate possible differences in mean particle size, PSD, purity.
	• Set up experimental program with appropriate parameters, including impeller type, speed, seed amount, seed size, rate of cooling or addition, point of addition.

- Analyze these results to determine the controlling steps, for example:
  - If the impeller type and speed has a small effect the nucleation rate may be low and surface incorporation may be controlling growth.
  - If some or all of these variables have a large effect, nucleation/growth rates may be high; the point of addition effect on mean particle size and PSD may be the most sensitive indicator of a high nucleation rate.
- Determine effect of impurities on nucleation and growth by spiking with process materials.

Analysis of results for effect on chemical purity

- If chemical purity is affected, this suggests rapid nucleation/growth leading to occlusion of impurities and/or solvent.
- If one or more impurities is unaffected by these variables, this suggests that the surface incorporation step is controlling for that impurity.
- If spiking with impurities changes the growth rate and/or the morphology, this suggests an effect of impurity concentration. If one or more impurities is consistently present at the same amount, this suggests solid solutions.

Note:

- Changes in hydration/solvation may indicate that the crystallization conditions are passing through different regions of equilibrium composition, requiring determination of these regions of stability for the desired product
- Test by changing the concentration of the crystallizer feed; the distribution of an impurity that forms a solid solution is proportional to the concentration in the liquor

Analysis of results for physical attributes

- If the mean particle size and PSD are not affected by these variables, the crystallization may be controlled by the surface incorporation/growth rate—also indicating a low nucleation rate and/or long induction time.
- If the mean particle size is small, this indicates excessive nucleation and/or a low growth rate.
- If mean particle size depends strongly on the point of addition, this indicates a high nucleation rate causing sensitivity to variability in local supersaturation at the different addition points.
- If the distribution is bimodal, this, indicates (1) that the addition or cooling rate is too high, causing additional nucleation after some growth or (2) that there is breakdown of agglomerates.
- If agglomerates occur, there is possible high global or local supersaturation causing rapid nucleation and agglomeration.

- If there is a change in morphology, this is a possible effect of impurities or high concentration gradients in the film and at the growing surface.

Note: Physical attributes are a strong function of the natural morphology and can be manipulated to some extent, as indicated below under corrective actions

Corrective actions to achieve growth	• Determine growth potential in a purified system by seeding at low supersaturation with low or no mixing.
	• If the crystals produced are sufficiently large (>20 microns) and three dimensional, they can be used as seeds.
	• If the crystals produced are high aspect ratio needles or flat plates, seed crystals should be prepared separately, as discussed in Section 5.5.

Note: Preferred morphology for growth, good filtration, and, downstream qualities, three-dimensional; whether the basic shape is a needle, plate, or cube.

- Run crystallization experiments to determine the seed quantity and addition time or cooling rate to achieve growth.
- Seed with minimum 10% three-dimensional seeds from above and run experiments at two or three cooling or addition rates and two mixing speeds.
- Measure mean particle size and PSD of seeds and final crystals.
- If there is an increase in number with no increase in particle size, this indicates nucleation over growth.
- If there is approximately the same number of particles with a $2-3\times$ increase in size, this indicates growth over nucleation.
- Increase the addition time until there is no further change in particle size, no bimodal distribution, and no change in number of final crystals.
- Test high and low agitator speeds to determine the effect on the number of final crystals, either by increased nucleation or crystal fracture.
- Perform further experimentation with amount and size of seed, agitator speed, and time of addition to optimize growth conditions and meet requirements.
- Determine feasibility for filtration, washing, and drying.

In-line methods for minimum particle size	Establish feasibility of nucleation rate (high), induction time (short), reaction rate (high) for short contact time nucleation.
Impinging jets for antisolvent or reactive crystallization	Optimize jet velocity to achieve required mean particle size and PSD.

Scale-up	Pilot plant/manufacturing
Development laboratory-scale experiment results can be used for scale-up	• Pilot plant vessel to include variable-speed drive, subsurface addition at impeller, online measurement devices, variable-temperature jacket services, controlled seed addition as slurry, antisolvent/reagent addition at controlled rate. • Maintain impeller type, batch height/diameter ratio, baffles. • Minimum recommended agitator speed—to achieve slurry suspension. • Maximum recommended agitator speed—equal power/volume (P/V).

Test in pilot plant	From pilot plant results, determine variables—agitator speed, cooling/addition rate, seed amount and timing, effect of point of addition—which contribute to differences in chemical purity and physical attributes.
	
	• Seed
	  • Amount: calculate amount required based on seed mean particle size. If seed mean particle size is too small, grow seed in the laboratory as needed (Chapter 5).
	  • Seeding point: determine method/timing of seeding as suggested in Chapter 5 and from observation. Maintaining amount, size, timing of seed addition, and point of addition, test as follows:
	
	• Agitator speed
	  • Increase until mean particle size decreases and width of PSD increases.
	  • If no limit is reached, decrease until slurry suspension is inadequate.
	  • Select speed for scale-up to achieve adequate suspension with desired attributes (Chapter 6).
	• Cooling rate
	  • Test natural cooling versus programmed cooling.
	  • If there is no wall scale or decrease in mean particle size with natural cooling, the system may be satisfactory provided that manufacturing uses the same temperature coolant.
	  • If there are issues with any aspect of the above, suggest programmed cooling and control of cooling fluid temperature (Chapter 7).
	• Addition rate for antisolvent or reagent
	  • Test linear versus programmed addition rate.
	  • If there is no decrease in mean particle size from seed with the linear rate, the system may be satisfactory but a slower addition rate may still be necessary on scale-up.

- • If programmed addition and longer addition time increase mean particle size or decrease impurities, design for use in manufacturing.
- • Extend addition time until no further increase in mean particle size occurs (Chapters 9 and 10).
- • Point of addition of antisolvent or reagent
    - • After determining a satisfactory combination of the above variables, test above-surface addition versus impeller discharge addition.
    - • If there is no change, can possibly use above surface addition on scale-up—but exercise care.
    - • If there is any change in by-product composition or physical attributes, use subsurface addition (Chapters 5, 6, 9, and 10).

Scale-up to manufacturing

- • Agitator speed: calculate speed based on speed determined in pilot plant using volume ratio, impeller diameter in power correlation.
- • Cooling rate: use the natural or programmed rate as determined above. (Use care in choosing the cooling medium temperature to avoid wall scale and wall nucleation.) Provide long exposure times to foreign particles.
- • Addition rate of antisolvent or reagent—use linear or programmed addition rate as determined above, decreasing the rate by a factor depending on the volume ratio.
- • Point of addition: use subsurface addition at the impeller unless pilot plant runs using above surface addition indicate no sensitivity with minimum agitation (prove that subsurface addition is not needed).

References

ALLEGRETTI, J.E. and M. MIDLER (1967). Continuous freezing process for purifying dimethyl sulfoxide. U.S. Patent 3,358,037.

ALLEGRETTI, J.E. and M. MIDLER (1970). Separation of stereoisomers by fluidized bed crystallization. Presented at AIChE annual meeting, Cleveland, OH, November.

ALSENZ, J. and M. KANSY (2007). High throughput solubility measurement in drug discovery and development. *Advanced Drug Delivery Reviews* **59**(7), 546–567.

AM ENDE, D., T.C. CRAWFORD, and N.P. WESTON (2003). Reactive crystallization method to improve particle size. U.S. Patent 6,558,436.

ANDERSON, H.W., J.B. CARBERRY, H.F. SATUNTON, and B.C. SUTRADHAR (1995). Crystallization of adipic acid. U.S. Patent 5,471,001.

ANGUS, J.C. and C.C. HAYMAN (1988). Low-pressure, metastable growth of diamond and "diamondlike" phases. *Science* **241**, 913.

ÅSLUND, B.L. and A.C. RASMUSON (1992). Semibatch reaction crystallization of benzoic acid. *AIChE J.* **38**(3), 328–342.

BALDYGA, J. and J.R. BOURNE (1999). *Turbulent Mixing and Chemical Reactions*. John Wiley & Sons, Chichester, UK.

BALZHISER, R.E., M.R. SAMUELS, and J.D. ELIASSEN (1972). *Chemical Engineering Thermodynamics*. Prentice-Hall, Englewood Cliffs, NJ.

BENET, N., L. FALK, H. MUHR, and E. PLASARI (1999). Experimental study of a two-impinging-jet mixing device for application in precipitation processes. *Int. Symp. Ind. Cryst. 14th* (computer optical disc), 1007–1016 (IChemE, Rugby, UK).

BENNEMA, P. (1969). The influence of surface diffusion for crystal growth from solution. *J. Cryst. Growth* **5**, 29–43.

BIRCH, M., S. FUSSELL, P.D. HIGGINSON, N. MCDOWELL, and I. MARZIAN (2005). Towards a PAT-based strategy for crystallization development. *Org. Process Res. Dev.* **9**, 360–364.

BOLZERN, O. and J.R. BOURNE (1985). Rapid chemical reactions in a centrifugal pump. *Chem. Eng. Res. Des.* **63**, 275–282.

BONNET, P.E., K.J. CARPENTER, and R.J. DAVEY (2002). A study into the phenomenon of oiling out.

Proceedings of the 15th International Symposium on Industrial Crystallization, p. 35.

BORNSTEIN, M. and M.D. CISE (1979). Method of preparing a rapidly dissolving powder of sterile crystalline cefazolin sodium for parenteral administration. U.S. Patent 4,146,971.

BOURNE, J.R. and J. GARCIA-ROSAS (1984). Rapid mixing and reaction in a Y-shape in-line dynamic mixer. *Paper 52c*, presented at AIChE annual meeting, San Francisco, November.

BOURNE, J.R. and J. GARCIA-ROSAS (1986). Rotor stator mixers for rapid micromixing. *Chem. Eng. Res. Des.* **64**, 11–17.

BRITTAN, H. (1999). *Polymorphism in Pharmaceutical Solids*. Marcel Dekker, New York.

BURGER, A. and R. RAMBERGER (1979). On the polymorphism of pharmaceuticals and other molecular crystals. II: Applicability of thermodynamic rules. *Mikrochimica Acta* **1979**, 259–316.

BURTON, W.K., N. CABRERA, and F.C. FRANK (1951). The growth of crystals and the equilibrium structure of their surfaces. *Philos. Trans.* **A243**, 299–358.

BYRN, S.R. (1982). *Solid-State Chemistry of Drugs*, Academic Press, New York.

CASEY, J.T. and A.I. LIAPIS (1985). Fixed bed sorption with recycle. Part III: Consecutive reversible reactions. *Chem. Eng. Res. Des.* **63**, 398.

Chemical Engineering magazine staff (1965). Kirkpatrick Chemical Engineering Achievement Awards. *Chem. Eng.* **72**(23), 247.

CHERNOV, A.A. (1961). The spiral growth of crystals. *Soviet Physics Uspekhi* **4**, 116–148.

CHOUDHURY, N.H., W.R. PENNEY, K. MEYERS, and J.B. FASANO (1995). An experimental investigation of solids suspension at high solids loadings in mechanically agitated vessels. *AIChE Symp. Ser.* **305**(91), 131–138.

CISE, M.D. and M.L. ROY (1979). Method of preparing a rapidly dissolving powder of crystalline cephalothin sodium for parenteral administration. U.S. Patent 4,132,848.

CLAS, S.D. (2003). The importance of characterizing the crystal form of the drug substance during drug development. *Current Opinion in Drug Discovery and Development* **6**, 550–560.

CLONTZ, N.A. and W.L. MCCABE (1971). Contact nucleation of $MgSO_4 \cdot 7H_2O$. *Chem Eng Prog Symp Series* **67**(110), 6–17.

CONDON, J.M. (2001). Investigation of impinging jet crystallization for a calcium oxalate model system. Ph.D. dissertation, Rutgers University, New Brunswick, NJ.

CONNOLLY, M., P. DEBENEDETTI, and H.H. TUNG (1996). Freeze crystallization of imipenem. *J. Pharm. Sci.* **85**, 174–177.

CROCKER, L.S. and J.A. MCCAULEY (1995). Comparison of the crystallinity of imipenem samples by X-ray diffraction of amorphous material. *J. Pharm. Sci.* **84**, 226–227.

CROCKER, L.S. and J.A. MCCAULEY (1997). Solubilities of losartan polymorph. *Pharmazie* **52**, 72.

DAVEY, R.J. and J. GARSIDE (2000). *From Molecules to Crystallizers: An Introduction to Crystallization.* Oxford University Press, Oxford, England.

DAVEY, R.J., S.J. JANCIE, and E.J. DE JONG (eds.) (1982). The role of additives in precipitation processes. *Industrial Crystallization 81* (8th Symposium, Budapest), North-Holland, Amsterdam, 123–135.

DAVEY, R.J., L.A. POLYWKA, and S.J. MAGINN (1991). The control of morphology by additives: molecular recognition, kinetics and technology. In *Advances in Industrial Crystallization*, J. Garside, R.J. Davey, and A.G. Jones (eds.). Butterworth-Heinemann, Oxford, 150–165.

DEBENEDETTI, P.G. (1995). *Metastable Liquids—Concepts and Principles.* Princeton University Press, Princeton, NJ.

DELUCA, P. and L. LACHMAN (1965). Lyophilization of pharmaceuticals. I. Effect of certain physical-chemical properties. *J. Pharm. Sci.* **54**, 617–624.

DEMYANOVICH, R.J. and J.R. BOURNE (1989). Rapid micromixing by the impingement of thin liquid sheets. *Ind. Eng. Chem. Res.* **28**, 825–830.

DENEAU, E. and G. STEELE (2005). An in-line study of oiling out and crystallization. *Organic Process R & D* **9**, 943–950.

DOLLING, U.H., ET AL. (1999). Finasteride processes. U.S. Patent 5,886,184.

DUNUWILA, D.D. and K.A. BERGLUND (1997). ATR-FTIR spectroscopy for in-situ measurement of supersaturation. *J. Cryst. Growth* **179**, 185–193.

EDWARDS, M.P.H. (1984). Chemical reaction engineering of polymer processing. Presented at ISCRE 8, Scotland, September.

ELDER, J. (1988). The thermal behavior of lovastatin. *Thermochimica Acta* **134**, 41–47.

ELDER, J. (1990). A new accelerated oxidative stability test for glass-forming organic compounds. *Thermochimica Acta* **166**, 199–206.

FINDLAY, A., A.N. CAMPBELL, and N.O. SMITH (1951). *The Phase Rule*, 9th ed. Dover Publications, New York, 35–42.

FRANK, F.C. (1949). The influence of dislocations on crystal growth. *Disc. Faraday Soc.* **5**, 48–54.

FRANK, T.C., J.R. DOWNEY, and S.K. GUPTA (1999). Quickly screen solvents for organic solids. *Chem. Eng. Prog.* **95**(12), 41–61.

FRANKS, F. (1990). Freeze drying: From empiricism to predictability. *Cryo-Letters* **11**, 93–110.

FUJIWARA, M., Z.K. NAGY, J.W. CHEW, and R.D. BRAATZ (2005). First-principles and direct design approaches for the control of pharmaceutical crystallization. *J. Process Control* **15**, 493–504.

GARSIDE, J. (1971). The concept of effectiveness factors in crystal growth. *Chem. Eng. Sci.* **26**, 1425–1431.

GARSIDE, J. and N.S. TAVARE (1981). Non-isothermal effectiveness factors for crystal growth. *Chem. Eng. Sci.* **36**, 863–866.

GARSIDE, J. and N.S. TAVARE (1985). Mixing, reaction, and precipitation: limits of micromixing in an MSMPR crystallizer. *Chem. Eng. Sci.* **40**, 1485–1493.

GATLIN, L. and P. DELUCA (1980). Study of phase transitions in frozen antibiotic solutions by DSC. *J. Parenteral Drug Assoc.* **34**, 398–408.

GENCK, W.J. (2003). Optimizing crystallizer scaleup. *Chem. Eng. Prog.* **99**(6), 36–44.

GILMER, G.H., R. GHEZ, and N. CABRERA (1971). An analysis of combined surface and volume diffusion processes in crystal growth. *J. Cryst. Growth* **8**, 79–93.

GIRON, D. (2000). Characterization of pharmaceuticals by thermal analysis. *Am. Pharm. Rev.* **3**, 53–61.

GRIFFITHS, H. (1925). Mechanical crystallization. *J. Soc. Chem. Ind.* **44**, 7T–18T.

GUPTA, R.B. (2006). Supercritical fluid technology for particle engineering. In *Nanoparticle Technology for Drug Delivery*, R.B. Gupta and U.B. Kompella (eds.). Informa Healthcare, New York, 53–84.

HACHERL (CONDON), J.M. (2003). Investigation of impinging-jet crystallization with a calcium oxalate model system. *AIChE J.* **49**, 2352–2362.

HALEBLIAN, J. and W. MCCRONE (1969). Pharmaceutical applications of polymorphism. *J. Pharm. Sci.* **58**, 911.

HANCOCK, B. and G. ZOGRAFI (1997). Characteristics and significance of the amorphous state in pharmaceutical systems. *J. Pharm. Sci.* **86**, 1–12.

HARNBY, N., M.F. EDWARDS, and A.W. NIENOW (eds.) (1992). *Mixing in the Process Industries*, 2nd ed. Butterworth, Stoneham, MA.

HARRISON, T., A.P. OWENS, B.J. WILLIAMS, C.J. SWAIN, A. WILLIAMS, E.J. CARLSON, W. RYCROFT, F.D. TATTERSALL, M.A. CASCIERI, G.G. CHICCHI, S. SADOWSKI, N.M.J. RUPNIAK, and R.J. HARGREAVES (2001). An orally active, water-soluble neurokinin-1 receptor antagonist suitable for both intravenous and oral clinical administration. *J. Med. Chem.* **44**, 4296.

HEM, S.L. (1967). The effects of ultrasonic vibrations on crystallization processes. *Ultrasonics* **5**, 202–207.

HOBBS, D.M., P. SCHUBERT, and H.H. TUNG (1997). Applying solubility to the design of reaction systems. *Ind. Eng. Chem. Res.* **36**, 302.

HOUCINE, I., E. PLASARI, R. DAVID, and J. VILLERMAUX (1997). Influence of mixing characteristics on the quality and size of precipitated calcium oxalate in a pilot scale reactor. *Trans IChemE* **75**, Part A, 252–256.

IP, D.P., G.S. BRENNER, J.M. STEVENSON, S. LINDENBAUM, A.W. DOUGLAS, S.D. KLEIN, and J.A. McCAULEY (1986). High resolution spectroscopic evidence and solution calorimetry studies on the polymorphs of enalapril maleate. *Int. J. Pharm.* **28**, 183–191.

ITO, K., T. AKOSHI, and S. TATSUMI (1966). Method of optically resolving racemic amino acids. U.S. Patent 3,260,744.

JENKINS, R. (2000). Use of x-ray powder diffraction in the pharmaceutical industry. *Am. Pharm. Rev.* **3**, 36–40.

JO, M.C., W.R. PENNEY, and J.B. FASANO (1994). Backmixing into reactor feed pipes caused by turbulence in an agitated vessel. *AIChE Symp. Ser. 299*, **90**, 41–49.

JOHNSON, B.J. (2003). Flash nanoprecipitation of organic actives via confined micromixing and block copolymer stabilization. Ph.D. dissertation, Princeton University, Princeton, NJ.

JOHNSON, B.J. and R.J. PRUD'HOMME (2003). Chemical processing and micromixing in confined impinging jets. *AIChE J.* **49**, 2264–2282.

JOHNSON, B.K., C. SZETO, O.A. DAVIDSON, and A.T. ANDREWS (1997). Optimization of pharmaceutical batch crystallization for filtration and scale-up. Presented at AIChE annual meeting, November.

JONES, A.G. (1974). Optimal operation of a batch crystallizer. *Chem. Eng. Sci.* **29**, 1075–1087.

JONES, A.G. and J.W. MULLIN (1974). Programmed cooling crystallization of potassium sulfate solutions. *Chem. Eng. Sci.* **29**, 105–118.

JUNG, J. and M. PERRUT (2001). Particle design using supercritical fluids: Literature and patent survey. *J. Supercritical Fluids* **20**, 179–219.

KHANKARI, R.K. and D.J.W. GRANT (1995). Pharmaceutical hydrates. *Thermochimica Acta* **248**, 62–79.

KIM, K.J. and A. MERSMANN (2001). Estimation of metastable zone width in different nucleation processes. *Chem. Eng. Sci.* **56**, 2315–2324.

KIM, S., B. LOTZ, M. LINDRUD, K. GIRARD, T. MOORE, K. NAGARAJAN, M. ALVAREZ, T. LEE, F. KIKFAR, M. DAVIDOVICH, S. SRIVASTAVA, and S. KIANG (2005). Control of the particle properties of a drug substance by crystallization engineering and the effect on drug product formulation. *Organic Process R & D* **9**, 894–901.

KING, A.O., E.G. CORLEY, R.K. ANDERSON, R.D. LARSEN, T.R. VERHOEVEN, and P.J. REIDER (1993). An efficient synthesis of LTD4 antagonist L-699392. *J. Org. Chem.* **58**, 3731–3735.

KLINK, A., M. MIDLER, and J.E. ALLEGRETTI (1971). A study of crystal cleavage by sonifier action. *Chem. Eng. Progress Symposium Series* **67**, No. 109 (*Sonochemical Engineering*), 74–80.

KOLAR, P., J.W. SHEN, A. TSUBOI, and T. ISHIKAWA (2002). Solvent selection for pharmaceuticals. *Fluid Phase Equilibrium* **194–197**, 771–782.

KRIEGER, K.H., J. LAGO, and J.A. WANTUCK (1968). Process for resolving racemic mixtures of optically-active enantiomorphs. U.S. Patent 3,405,159.

KUHNERT-BRANDSTATER, M. (1971). *Thermomicroscopy in the Analysis of Pharmaceuticals.* Pergamon Press, New York, p. 42.

LAFFERRERE, L., C. HOFF, and S. VEESLER (2002). Polymorphism and liquid–liquid phase separation of pharmaceutical compounds. *Proceedings of the 15th International Symposium on Industrial Crystallization*, p. 819.

LAGO, J., K.H. KRIEGER, and J.A. WANTUCK (1966). Development and design of a process for the resolution of *dl*-alpha-methyl-3,4-dihydroxyphenylaniline. Presented at AIChE Meeting, Columbus, OH, November.

LARSON, K.A., M. MIDLER, and E.L. PAUL (1995). Reactive crystallization: control of particle size and scale-up. Presented at the Association for Crystallization Technology meeting, Charlottesville, VA, March.

LAUFHUTTE, H.D. and A. MERSMANN (1987). Local energy dissipation in agitated turbulent fluids and its significance for the design of stirring equipment. *Chem. Eng. Technol.* **10**, 56–63.

LEE, L.J., J.M. OTTINO, E.E. RANZ, and C.W MACOSKO (1980). Impingement mixing in reaction injection molding. *Polymer Eng. & Sci.* **20**(13), 868–874.

LEUNER, C. and J. DRESSMAN (2000). Improving drug solubility for oral delivery using solid dispersion. *Euro. J. Pharm. BioPharm.* **50**, 47–60.

LEVENSPIEL, O. (1972). *Chemical Reaction Engineering.* John Wiley & Sons, New York.

LINDBERG, M. and A.C. RASMUSON (2000). Supersaturation generation at the feed point in reaction crystallization of a molecular compound. *Chem. Eng. Sci.* **55**, 1735–1746.

LINDRUD, M.D., S. KIM, and C. WEI (2001). Sonic impinging jet crystallization apparatus and process. U.S. Patent 6,302,958.

LIPINSKI, C.A., F. LOMBARDO, B.W. DOMINY, and P.J. FEENEY (2001). Experimental and computational approaches to estimate solubility and permeability in drug discovery and development settings. *Adv. Drug Delivery Rev.* **46**, 3–26.

LOWINGER, M., H.H. TUNG, H.C. McKELVEY, Z. LIU, and W. WU (2007). Investigating molecular interactions between drug and polymer molecules in solid amorphous dispersions. Presented at the *Colloid and Surface Science 81st Symposium*, Newark, DE, June.

MacKENZIE, A.P. (1965). Factors affecting the mechanism of transformation of ice into water vapor in the freeze-drying. *Ann. NY Acad. Sci.* **125**, 522–547.

MacKenzie, A.P. (1977). The physico-chemical basis for the freeze-drying process. *Dev. Biol. Standard.* **36**, 51–67.

MacKenzie, A.P. (1985). *Refrigeration Science and Technology.* International Institute of Refrigeration, Paris, 21–34, 155–163.

Mahajan, A.J. (1993). Rapid precipitation of biochemicals, kinetics and micromixing. Ph.D. thesis, University of Virginia, Charlottesville, VA.

Mahajan, A.J. and D.J. Kirwan (1996). Micromixing effects in a two-impinging-jet precipitator. *AIChE J.* **42**, 1801–1814.

Manth, T., D. Mignon, and H. Offermann (1996). Experimental investigation of precipitation reactions under homogeneous mixing conditions. *Chem. Eng. Sci.* **51**, 2571–2576.

Marcant, B.N. and R. David (1991). Experimental evidence for and prediction of micromixing effects in precipitation. *AIChE J.* **37**, 1698–1710.

Martin, P., E.J. Phillips, and C.J. Price (1993). Power ultrasound—a new tool for controlling crystallization. *Proceedings of the 1993 IChemE Research Event*, Birmingham University, Birmingham, UK, 6–7 January.

McCabe, W.L. (1929). Crystal growth in aqueous solutions. *Ind. Eng. Chem.* **21**, 30–33, 112–119.

McCauley, J.A. (1991). Particle design via crystallization. *AIChE Symp. Ser.* **87**, 58.

McCauley, J.A., R.J. Varsolona, and D.A. Levorse (1993). The effect of polymorphism and metastability on the characterization and isolation of two pharmaceutical compounds. *J. Phys. D: Appl. Phys.* **26**, 85.

McCausland, L.J., P.W. Cains, and P.D. Martin (2001). Use the power of sonocrystallization for improved properties. *Chem. Eng. Prog.* **97**, 56.

Mersmann, A. (2001). *Crystallization Technology Handbook*, 2nd ed. Marcel Dekker, New York.

Mersmann, A. and M. Kind (1988). Chemical engineering aspects of precipitation from solution. *Chem. Eng. Technol.* **11**, 264–276.

Midler, M. (1970). Production of crystals in a fluidized bed with ultrasonic vibrations. U.S. Patent 3,510,266.

Midler, M. (1975). Process for production of crystals in fluidized bed crystallizers. U.S. Patent 3,892,539.

Midler, M. (1976). Crystallization system and method using crystal fracturing external to a crystallizer column. U.S. Patent 3,996,018.

Midler, M., E. Paul, E. Whittington, M. Futran, P. Liu, J. Hsu, and S. Pan (1994). Crystallization method to improve crystal structure and size. U.S. Patent 5,314,506.

Moore, M.L. (1943). U.S. Patent, 2334015.

Mullin, J.W. (1993). *Crystallization*, 3rd ed., Butterworth-Heinemann, Oxford.

Mullin, J.W. (2001). *Crystallization*, 4th ed. Butterworth-Heinemann, Oxford.

Mullin, J.W. and J. Nyvlt (1971). Programmed cooling of batch crystallizers. *Chem. Eng. Sci.* **26**, 369–377.

Myerson, A.S. (1999). *Molecular Modelling Applications in Crystallization.* Cambridge University Press, Cambridge.

Myerson, A.S. (ed.) (2002). *Handbook of Industrial Crystallization*, 2nd ed. Butterworth-Heinemann, Newton, MA.

Nagy, Z.K., M. Fujiwara, and R.D. Braatz (2007). Recent advances in the modeling and control of cooling and antisolvent crystallization of pharmaceuticals. *Proceedings of the 8th International IFAC Symposium on Dynamics and Control of Process Systems* **2**, 29.

Ness, J.N. and E.T. White (1976). Collision nucleation in an agitated crystallizer. *AIChE Symposium Series* **72**(153), 64–73.

Nielsen, A.E. (1964). *Kinetics of Precipitation.* Pergamon Press, New York.

Nienow, A.W. (1976). The effect of agitation and scale-up on crystal growth rates and on secondary nucleation. *Trans. Inst. Chem. Eng.* **54**, 205.

Nienow, A.W. and K. Inoue (1993). A study of precipitation micromixing, macromixing, size distribution, and morphology. Paper 9.4, presented at CHISA, Prague, Czech Republic.

Nonoyama, N., K. Hanaki, and Y. Yabuki (2006). Constant supersaturation control of anti-solvent-addition batch crystallization. *Org. Process R & D* **10**, 727–732.

Nyvlt, J. (1971). *Industrial Crystallization from Solution.* Butterworth, London.

Nyvlt, J., H. Kocova, and M. Cerny (1973). Size distribution of crystals from a batch crystallizer. *Coll. Czech. Comm.* **38**, 3199–3209.

Nyvlt, J., O. Söhnel, M. Matuchová, and M. Broul (1985). *The Kinetics of Industrial Crystallization.* Academia, Prague.

Oberholtzer, E.R. (1988). *Analytical Profiles of Drug Substances*, Vol. 17. Academic Press, New York, pp. 73–144.

Ohara, M. and R.C. Reid (1973). *Modeling Crystal Growth Rates from Solution.* Prentice-Hall, Inc., Englewood Cliffs, NJ.

O'Sullivan, B. and B. Glennon (2006). Application of in-situ FBRM and ATR-FTIR to the monitoring of the polymorphic transformation of D-mannitol. *Organic Process R & D* **9**, 884–889.

Ostwald, W. (1897). Studien uber die Bildung und Umwandlung fester Korper. *Z. physik. Chem.* **22**, 289.

Paul, E.L. (1990). Reaction systems for bulk pharmaceutical production. *Chem. Ind.* **21**, 320–325.

Paul, E.L., V. Atiemo-Obeng, and S.M. Kresta (eds.) (2003). *Handbook of Industrial Mixing: Science and Practice.* John Wiley & Sons, New York.

PAUL, E.L., H.H. TUNG, and M. MIDLER (2006). Organic crystallization process. *Powder Technol.* **150**, 133–143.

PESSLER, R. (1997). Batch crystallization. *Pharm. Eng.* **17**, 42–46.

PHILLIPS, R., S. ROHANI, and J. BALDGYA (1999). Micromixing in a single-feed semi-batch precipitation process. *AIChE J.* **45**, 82–92.

PIKAL, M. (1990). Freeze-drying of proteins, Part I: Process design. *Biopharm.* **3**(8), 18–27.

PODKULSKI, D.E. (1997). How do new process analyzers measure up? *Chem. Eng. Prog.* **93**(10), 33–46.

POHARECKI, R. and J. BALDGYA (1988). The effects of micromixing and the manner of reactant feeding on precipitation in stirred tank reactors. *Chem. Eng. Sci.* **43**, 1949–1954.

PRICE, C.J. (1997a). Take some solid steps to improve crystallization. *Chem. Eng. Prog.* **93**(9), 34.

PRICE, C.J. (1997b). Ultrasound—the key to better crystals for the pharmaceutical industry. *Pharm. Technol. Eur.* **9**(10), 78.

RAGHAVAN, R., A. DWIVEDI, G.C. CAMPBELL, JR., E. JOHNSTON, D. LEVORSE, J. McCAULEY, and M. HUSSAIN (1993). *Pharm. Res.* **10**, 900.

RANDOLPH, A.D. and M.A. LARSON (1971). *Theory of Particulate Processes.* Academic Press, New York.

RANDOLPH, A.D. and M.A. LARSON (1988). *Theory of Particulate Processes,* 2nd Academic Press, San Diego, CA.

RASMUSSON, G.H., G.F. REYNOLDS, N.G. STEINBERG, E. WALTON, G.F. PATEL, T. LIANG, M.A. CASCIERI, A.H. CHEUNG, J.R. BROOKS, and C. BERMAN (1986). Azasteroids: Structure-activity relationships for inhibition of 5 alpha-reductase and of androgen receptor binding. *J. Med. Chem.* **29**, 2298.

REID, R.C., J.M. PRAUSNITZ, and T.K. SHERWOOD (1977). *The Properties of Gases and Liquids,* 3rd ed. McGraw-Hill Book Company, New York.

ROSSO, J.C., C. CANALS, and L. CARBONNEL (1975). Formation of water-acetone clathrate. *C.R. Acad. Sc. Paris, Series C* **281**, 699–702.

ROSAS, C.B. (1997). Personal communication.

ROUHI, A.M. (2003). The right stuff. *Chem. Eng. News* **81**(8), 32–35.

RUECROFT, G., D. HIPKISS, T. LY, N. MAXTED, and P. CAINS (2006). Sonocrystaliztion: the use of ultrasound for improved industrial crystallization. *Organic Process R & D* **9**, 923–932.

RUTHVEN, D.M. (1984). *Principles of Adsorption and Adsorption Processes,* Wiley, New York.

SARANTEAS, K., R. BAKALE, Y. HONG, H. LUONG, R. FOROUGHI, and S. WALD (2005). Process design and scale-up elements for solvent mediated polymorphic controlled tecastemizole crystallization. *Organic Process R & D* **9**, 911–922.

SARETT, L. (1956). Delta1,4-3,20-diketo-11-oxygenated-17,21-dihydroxy-pregnadiene 21-tertiary

butyl acetates and 9-fluoro derivatives thereof. U.S. Patent 2,736,734.

SCHUBERT, H. and A. MERSMANN (1996). Determination of heterogeneous nucleation rates. *Chem. Eng. Res. Des.* **74**, 821–827.

SHARMA, M.M. (1988). Multiphase reactions in the manufacture of fine chemicals. *Chem. Eng. Sci.* **43**, 1749–1750.

SHEKUNOV, B.Y., J. BALDGYA, and P. YORK (2001). Particle formation by mixing with supercritical antisolvent at high Reynolds numbers. *Chem. Eng. Sci.* **56**(7), 2421–2433.

SHELDON, R.A. (1993). *Chirotechnology.* Marcel Dekker, New York, NY.

SHEN, T.Y., T.B. WINDHOLZ, A. ROSEGAY, B.E. WITZEL, A.N. WILSON, J.D. WILLETT, W.J. HOLTZ, R.L. ELLIS, A.R. MATZUK, S. LUCAS, C.H. STAMMER, F.W. HOLLY, L.H. SARETT, E.A. RISLEY, G.W. NUSS, and C.A. WINTER (1963). Nonsterioid antiinflammatory agents. *J. Am. Chem. Soc.* **85**, 488.

SHEN, T.Y., R.B. GREENWALD, H. JONES, B.O. LINN, and B.E. WITZEL (1972). Substituted indenyl acetic acids, US Patent 3,654,349.

SMOLUCHOWSKI, M. (1918). von Versuch einer mathematischen theorie der koagulations kinetic kolloider lösungen. *Z. physik. Chem.* **92**, 129–168.

SÖHNEL, O. and J. GARSIDE (1992). *Precipitation: Basic Principles and Industrial Applications.* Butterworth Heinemann, Oxford, p. 391.

SÖHNEL, O. and J. GARSIDE (1993). *Precipitation.* Butterworth Heinemann, Oxford.

STOIBER, R. and S. MORSE (1994). *Crystal Identification with the Polarizing Microscope.* Chapman and Hall, London and New York

SUBRAMANIAM, B., R.A. RAJEWSKI, and K. SNAVELY (1997). Pharmaceutical processing with supercritical carbon dioxide. *J. Pharm. Sci.* **86**, 885–890.

SUSLICK, K.S. (ed.) (1988). *Ultrasound—Its Chemical and Biological Effects.* VCH, New York/Weinheim.

SUSLICK, K.S., S.J. DOKTYCZ, and E.B. FLINT (1990). On the origin of sonoluminescence and sonochemistry. *Ultrasonics* **28**, 280.

TAVARE, N.S. and J. GARSIDE (1990). Simulation of reactive crystallization in a semi-batch crystallizer. *Trans. Inst. Chem. Eng.* **68**, 115–122.

THOMPSON, L.H. and L.K. DORAISWAMY (2000). The rate enhancing effect of ultrasound by inducing supersaturation in a solid-liquid system. *Chem. Eng. Sci.* **55**, 3085–3090.

TOGKALIDOU, T., R.D. BRAATZ, B.K. JOHNSON, O.A. DAVIDSON, and A.T. ANDREWS (2001). Experimental design and inferential modeling in pharmaceutical crystallization. *AIChE J.* **47**, 160–168.

TOGKALIDOU, T., H.H. TUNG, Y. SUN, A. ANDREWS, and R.D. BRAATZ (2002). Solution concentration prediction for pharmaceutical crystallization processes using robust chemometrics and ATR FTIR spectroscopy. *Org. Process Res. Dev.* **6**, 317–322.

TOGKALIDOU, T., H.H. TUNG, Y. SUN, A. ANDREWS, and R.D. BRAATZ (2004). Parameter estimation and optimization of a loosely bound aggregating pharmaceutical crystallization using in situ infrared and laser backscattering measurements. *Ind. Eng. Chem. Res.* **43**, 6168–6181.

TORBACKE, M. and A.C. RASMUSON (2001). Influence of different scales of mixing in reaction crystallization. *Chem. Eng. Sci.* **56**, 2459–2473.

TUNG, H.H., A. EPSTEIN, M. SOWA, and J. GRAU (1998). Impinging jet crystallization of pharmaceuticals. *Proceedings of the International Symposium on Industrial Crystallization*, ISIC 1998, p138, Tianjin, China.

TUNG, H.H., D.M. HOBBS, and E.L. PAUL (1992). Reaction system design with solid precipitation. Presented at AIChE annual meeting, November.

TUNG, H.H., J. TABORA, N. VARIANKAVAL, D. BAKKE, and C.-C. CHEN (2007). Prediction of pharmaceutical solubility via NRTL-SAC and COSMO-SAC. *J Pharm. Sci.* **4**, 1813–1820.

TUNG, H.H., L. WANG, S. PANMAI, and M.T. RIEBE (2008). Effects of energy on the formation of drug nanoparticles under supersaturation. Presented at *Particles 2008*, Orlando, FL, May.

VARIANKAVAL, N., C. LEE, J. XU, R. CALABRIA, N. TSOU, and R. BALL (2007). Water activity-mediate control of crystalline phases of an active pharmaceutical ingredient. *Org. Process Res. Dev.* **11**(2), 229–236.

VERMA, A.R. and P. KRISHNA (1966). *Polymorphism and Polytypism in Crystals*. John Wiley and Sons, New York, 7–66.

VIPPAGUNTA, S.R., H.G. BRITTAIN, and D.J.W. GRANT (2001). Crystalline solids. *Advanced Drug Delivery* **48**, 3–26.

VOLMER, M. (1939). *Kinetic der Phasenbildung*. Steinkopff, Leipzig.

WANG, L. and R.O. FOX (2004). Comparison of micromixing models for CFD simulation of nanoparticle formation. *AIChE J.* **50**, 2217.

WANG, Y., R. LOBRUTTO, R.W. WENSLOW, and I. SANTOS (2005). Eutectic composition of a chiral mixture containing a racemic compound. *Organic Process R & D* **9**, 670–676.

WANG, Y., R.M. WENSLOW, J.A. MCCAULEY, and L.S. CROCKER (2002). Polymorphic behavior of an NK1 receptor antagonist. *Int. J. Pharmaceutics* **243**, 147.

WEISSBUCH, I., L.J.W. SHIMON, E.M. LANDAU, R. POPOVITZ-BIRO, L. ADDADI, Z. BERKOVITCH-YELLIN, M. LAHAV, and L. LEISEROWITZ (1995). Understanding and control of nucleation, growth, habit, dissolution and structure of two- and three-dimensional crystals using 'tailor-made' auxiliaries. *Acta Cryst.* **B51**, 115–148.

WENSLOW, R.M., M.W. BAUM, K. HOOGSTEEN, J.A. MCCAULEY, and R.J. VARSOLONA (2000). A spectroscopic and crystallographic study of polymorphism in an aza-steroid. *J. Pharm. Sci.* **89**, 1271.

WINN, D. and M.F. DOHERTY (2002). Predicting the shape of organic crystals grown from polar solvents. *Chem. Eng. Sci.* **57**(10), 1805–1813.

WOO, X.Y., R.B.H. TAN, P.S. CHOW, and R.D. BRAATZ (2006). Simulation of mixing effects in antisolvent crystallization using a coupled CFD-PDF-PBE approach. *Crystal Growth & Design* **6**, 1291–1303.

YANG, S.S. and J.L. GUILLPRY (1972). Polymorphism in sulfonamides. *J. Pharm. Sci.* **61**, 26.

YORK, P. (1999). Strategies for particle design using supercritical fluid technologies. *Pharmaceutical Science & Technology Today (PSTT)* **2**, 430–440.

YORK, P., U.B. KOMPELLA, and B.Y. SHEKUNOV (eds.) (2004). *Supercritical Fluid Technology for Drug Product Development*. Informa Health Care, New York.

YOUNG, F.R. (1989), *Cavitation*. McGraw-Hill Book Company, London.

YU, L. (1995). Inferring thermodynamic stability relationship of polymorphs from melting data. *J. Pharm. Sci.* **84**, 966.

ZERNIKE, J. (1955). *Chemical Phase Theory*. Uitgevers Maatschappif, Antwerp, 429.

ZHOU, X., M. FUJIWARA, X.Y. WOO, E. RUSLI, H.H. TUNG, C. STARBUCK, O. DAVIDSON, Z.H. GE, and R.D. BRAATZ (2006). Direct design of pharmaceutical antisolvent crystallization through concentration control. *Crystal Growth & Design* **6**, 892–898.

ZLOKARNIK, M. (2001). *Stirring: Theory and Practice*. Wiley-VCH, Weinheim, Germany.

ZWIETERLING, T.N. (1958). Suspending of solid particles in liquid by agitators. *Chem. Eng. Sci.* **8**, 244.

Index

Crystallization of Organic Compounds: An Industrial Perspective. By H.-H. Tung, E. L. Paul, M. Midler, and
J. A. McCauley